UNION ARMY BALLOON CORPS

OPERATIONS *DURING THE* WAR *OF THE* REBELLION 1861-1863

Russell K. Dutcher, III
F.A.C.G.

HERITAGE BOOKS
2009

HERITAGE BOOKS
AN IMPRINT OF HERITAGE BOOKS, INC.

Books, CDs, and more—Worldwide

For our listing of thousands of titles see our website at
www.HeritageBooks.com

Published 2009 by
HERITAGE BOOKS, INC.
Publishing Division
100 Railroad Ave. #104
Westminster, Maryland 21157

Copyright © 2009 Russell K. Dutcher, III

Other books by the author:

Biographical Sketches of the Sheriffs of the County of Union in the State of New Jersey, 1857-1993: With Compiled Lists of the Constabulary of Some Select New Jersey Towns, 1669-1876

All rights reserved. No part of this book may be reproduced or transmitted in any form or by any means, electronic or mechanical, including photocopying, recording or by any information storage and retrieval system without written permission from the author, except for the inclusion of brief quotations in a review.

International Standard Book Numbers
Paperbound: 978-0-7884-4785-3
Clothbound: 978-0-7884-8062-1

Professor T.S.C. Lowe's Official Report

DEDICATION

The dedication of this book is simple,

To the

Memory of Professor T.S.C. Lowe

And the

Airmen of the

United States Air Force,

Our Defenders and Protectors

From Above

Professor T.S.C. Lowe's Official Report

TABLE OF CONTENTS

List of Illustrations -	vii
Foreword -	ix – x
Acknowledgements -	xii
Introduction –	xiv - xx
Biography of T.S.C. Lowe -	1 - 6
Lowe's Official Report – Part I	7 - 56
Illustrations -	57 - 63
Lowe's Official Report – Part II	64 - 172
Report Conclusion -	173 - 176
Lowe's Account of Operations During the Peninsula Campaign – 1862	177 - 183
Fitz-John Porter's View of the Confederates from a Balloon	184 - 187
Listing of the Balloon Fleet of the Aeronautic Corps	188

Professor T.S.C. Lowe's Official Report

LIST OF ILLUSTRATIONS

1. T.S.C. Lowe – 1861

2. Inflation of the Balloon - *Intrepid*

3. Twin Hydrogen Inflation Generators

4. Hydrogen Gas Generators at the U.S. Army Supply Park

5. Ascension of the *Intrepid* at the Battle of Fair Oakes, VA – 1862

6. Advertisement for the Balloon, *The City of New York*

7. T.S.C. Lowe

8. Leontine A.G. Lowe

9. Route of Lowe's Flight from Cincinnati, Ohio

10. Hydrogen Gas Generators at Work in the Field

11. The Note from President Abraham Lincoln Authorizing the Formation of the Union Army Balloon Corps

12. Schematics of Lowe's Hydrogen Gas Generators

13. Lowe's Balloon Corps Personnel in the Field and at the Ready

14. T.S.C. Lowe - 1862

Professor T.S.C. Lowe's Official Report

FOREWORD

While collecting the information necessary for compiling historical documentation regarding a factual account of early aviation history, I became quite fascinated with the innovations and work conducted by Professor Thaddeus Sobieski Constantine Lowe, during the late War of the Rebellion. He capably demonstrated how properly coordinated aerial operations could support an army in the field.

I would hope that this compilation of Lowe's Official Report to Congress will shed some light upon a man whose innovations have for many years remained overshadowed by more recent events in aviation history

It is quite appropriate that Lowe's many accomplishments be listed here and documented for future generations of military historians and aviation enthusiasts.

Under Lowes guidance, the Union Army Balloon Corps was able to conduct "real time" reconnaissance of enemy positions by utilizing air-to-ground communications via aerial telegraph. Additionally, Lowe was able to save valuable time regarding mobile field gas generating systems which allowed for inflation of the balloons in the field.

Previously, these behemoths needed to be filled at a gas refueling site and then brought by rail to the battlefield. Also, the use of incandescent lamps for night operations provided his corps of balloonists with the opportunity to repair and ready their balloons for the next day's activities.

Professor T.S.C. Lowe's Official Report

The consummate inventor and innovator, Lowe consistently strived to drive the idea of strategic aerial reconnaissance home to the Union Army General's Staff. He was rebuffed regarding most of his applications regarding enhanced aerial to ground operations.

However, General Fitz-John Porter, one of a small cadre of Staff Officers who supported Lowe and his Balloon Corps, lobbied tirelessly for increased financial and administrative support.

From it's inception in July 1861 throughout the end of December 1863, the Balloon Corps achieved admirable marked success regarding its support of the Army of the Potomac and its operations against a most formidable adversary, the Army of Northern Virginia.

With an unrelenting determination to venture into uncharted territories, he will most assuredly be known as the man who passed on a legacy of integrity, self-sacrifice and excellence, and hold undisputed claim to the title, *"FATHER OF THE UNITED STATES AIR FORCE."*

RUSSELL K. DUTCHER, III
Lakewood, New Jersey
November - 2008

Professor T.S.C. Lowe's Official Report

ACKNOWLEDGEMENTS

I wish to express my sincere appreciation to Gary Boyd, Historian, 305th Air Mobility Wing, McGuire Air Force Base, New Jersey, for his invaluable support during the research, organization, and presentation of the material contained within this book. I would also like to thank Lt Col Mark Ustaszewski, Safety Chief, 514th Air Mobility Wing, for his expertise and support of this project.

RUSSELL K. DUTCHER, III

Professor T.S.C. Lowe's Official Report

INTRODUCTION

Today, we look upon the Wright Brothers' flight at Kitty Hawk, North Carolina in 1903, as the premier event which sparked the beginning of American military aviation. To some extent this is true, but we must rewind some forty-two years previous to observe the true seeds of practical aeronautics which were sown during the War Between the States.

It can be stated categorically that the formation of the Union Army Balloon Corps during this conflict greatly changed not only aviation principles and practices, but greatly impacted the course of world history. Additionally, the use of balloons during the war forever changed the way in which future wars would be fought. This set the tone for other nations to follow and paved the way for modern strategic aerial warfare.

The first viable balloon was invented by the Montgolfier brothers in France. These men observed the principles and properties of heated air and then designed a lightweight vessel in which to capture the heated air. They then applied some basic principles of physics and set about attaching a device which would carry individuals in free flight. These efforts led to the first successful balloon demonstration on September 19, 1783.

Just three months after the first manned balloon flights in France in 1783, Benjamin Franklin wrote of the new invention's military capabilities. Franklin envisioned balloons carrying upwards of two to three men descending upon the enemy through surprise before their forces could be organized. His contention was that a few thousand balloons so equipped were less than the cost of "two ships of the line."

Professor T.S.C. Lowe's Official Report

These ascents were watched very carefully by members of the French military who understood the use of the balloon as a practical military device. By 1789, the French army had already made plans to integrate the balloon into their strategic planning doctrine.

Over the next few years, French military leaders intended to utilize the balloons for reconnaissance purposes and had appointed Captain J.M. Coutelle as chief of aeronautics. The French military balloonists (known as Aerostiers) were to be charged with identifying opposing military forces and directing their observations to their ground forces. The first full scale use of these observers was during the battle of Fleurus in 1794. The French balloonists provided critical information about troop movements, which enabled the French Generals to act quickly in positioning their forces. These ascents were very successful and greatly contributed to their victory.

As with Benjamin Franklin's observations, many Americans became interested in the art of hot air ballooning and on January 9, 1793, the first aerial ascension took place and was witnessed by President George Washington.

Members of the War Department may or may not have seen the usefulness of balloons regarding military operations. As stated earlier, the French were quick to see the value of aerial reconnaissance during ground operations, yet it would take the Americans over sixty years for the American Army to utilize balloons during wartime.

In the years leading up to the war, there were numerous aeronauts who were working to incorporate the balloon in military operations. In 1841, Colonel John Sherburne, of the United States Army, put forth the idea of utilizing the balloon for nighttime reconnaissance against the Seminole Indians. For some years, the United States had been involved in *on again-off again* military operations against the Seminoles. The Seminoles had to take refuge in the Florida everglades

Professor T.S.C. Lowe's Official Report

and had fought both Army and Marine forces to a standstill. Sherburne suggested that balloons be attached to head each American infantry column and then utilized during the evening hours to spot the camp fires of the Seminoles. This idea was sent to the War Department, but in the final review, was rejected by staff officers as being unmanageable.

Some five years later during the Mexican War, American Aeronaut, John Wise suggested the use of balloons against Mexican Forces, whereby explosive or incendiary devices would be unleashed over a fortified city and then triggered to drop over the target area.

This idea was also rejected due to the fact that the balloons could not be properly maneuvered from the ground and that a cumbersome trigger device upwards to five miles long would be required to release the explosive devices.

These ideas, though unworkable at the time, did have merit and would form the basis for future strategic aerial reconnaissance and bombing operations in the future.

Between the Mexican War and the start of the War Between the States, aeronauts were trying to find safer ways of ascending with their balloons, and the lighter than air hydrogen gas was now being utilized as the primary agent for lifting the balloons instead of hot air. This was in large part due to the fact that hot air balloons required the aeronaut to keep a fire burning. This was quite hazardous since burning embers from the fire could reach the balloon and cause a catastrophic and deadly fire.

When war broke out in April 1861, many of the leading aeronauts including John Wise, T.S.C. Lowe, James Allen and John LaMountain, saw the opportunity to utilize the balloons for military observation.

Professor T.S.C. Lowe's Official Report

Three of the leading aeronauts, Allen and Wise did have a head start on Lowe regarding the utilization of their balloons with the Union forces. James Allen was the first to get his balloons to the battlefield, but due to excessive damage done to the balloons during the transportation period, was unable to get them in the air.

Wise, was also vying for the opportunity of being the primary aeronaut for the Union Army, but lost the confidence of both the War Department and high ranking staff officers after he was unable to deliver his balloon to the government prior to the Battle of First Manassas in June 1861.

Professor T.S.C. Lowe was the main aeronaut who was competing with Wise for the government contract in Washington at that time. He understood the advantages which balloons could have on the battlefield. Lowe had gone to Washington, D.C. in hopes of garnering support for his program as well as an appointment as the Army's primary aeronaut. He had conducted balloon ascensions on the grounds of the Smithsonian Institution, which was in front of the White House.

Lowe had also gained support from President Abraham Lincoln regarding his experiments and then utilized the technology regarding the telegraph to send a message from the air to President Lincoln on the ground. This was the first successful air-to-ground conducted in history and the message to the President read,

Professor T.S.C. Lowe's Official Report

"Dear Sir"
"From this point of observation we command an extent of our country nearly fifty miles in diameter. I have the pleasure of sending you this first telegram ever dispatched from an aerial station, and acknowledging indebtedness to your encouragement for the opportunity of demonstrating the availability of the science of aeronautics in the service of the country."
"I am, Your Excellency's obedient servant,
T.S.C. Lowe"

On August 2 1861, President Lincoln sent a note to then General of the Army, Winfield Scott instructing him to utilize Lowe's services, thus signifying the beginning of the Union Army Balloon Corps.

With appointment in hand, Professor Lowe set about making the Aeronautic Corps battlefield ready along with the ability to accompany the Union Army throughout the upcoming campaigns. He continually looked for more innovative ways to get his balloons in the air and keep them there.

Lowe was able to institute many firsts regarding the aeronautic program. These included the use of mobile gas generators for real time inflation of the balloons in the field. Previous to this, balloons needed to be inflated at alternate locations and then transported by rail to the field for use.

He was also able to utilize air-to-ground communication regarding real-time reconnaissance regarding opposing enemy forces. This enabled Commanders to move their forces throughout the battlefield to meet additional threats.

Also, the balloons were able to direct accurate artillery fire on enemy positions due to their wide panoramic view of the field of battle.

Professor T.S.C. Lowe's Official Report

From August 1861 to December 1863, under the command of Professor T.S.C. Lowe, the Union Army Aeronautic Corps utilized six balloons which included: *Eagle, Constitution, Intrepid, Washington* and *Union*.

These balloons accompanied the Union Army during many of its campaigns and provided valuable support on various occasions. One in particular, occurred at the battle of Fair Oaks where the *Intrepid*, was launched and utilized by Lowe to observe enemy troops' movements. At one point, the balloon lost air, but through the ingenuity of Lowe, was able to be replenished with hydrogen which enabled it to stay aloft.

Through the sheer determination and heroism of Lowe, the information provided to the Union Commanders enabled maneuvering on the field which directly led to General Heintzelman's Army being destroyed due to isolation on the field.

Though there were many successes regarding the Aeronautic Corps, the continued disagreements between Lowe and his superiors regarding operational command and the clandestine attempts to undermine his authority, led to a reduction in force throughout late 1862 to 1863.

Lowe was stricken with malaria during the operations in Virginia and left the field to convalesce. During his absence, the Balloon Corps was not fully utilized and its equipment was pilfered. Also, the force necessary for the successful ascensions of the balloons was reduced and inexperienced men from infantry regiments were utilized during the ensuing year. This caused frustration and confusion with respect to the proper training needed to operate the balloons.

Diminished resources, government funding and a drastic reduction in pay caused a further deterioration of the Corps. By August 1863, due to this myriad of problems and limitations, the Balloon Corps ceased to exist as an effective unit.

Professor T.S.C. Lowe's Official Report

The Aeronautic Corps was later reassigned to the U.S. Army Engineer Corps, but was never again utilized to its fullest potential.

It can be said, however, that during its brief period of existence under direction of Professor T.S.C. Lowe, that many innovations were realized which would eventually pave the way for future advances within the field of the realm of aerial flight. Also, the men as well as the balloons of the corps not only secured for themselves a place in aeronautical history, but at the time, provided a sense of security for the military personnel which they supported.

The effect the balloons had on opposing enemy forces cannot be denied. Confederate forces admitted that the site of Union Army balloons was quite unnerving. The Confederate Army tried many times to bring down the Corps' balloons, but to no avail.

A *New York Times* article dated August 19, 1894, may aptly sum up the U.S. Army's support of the program, *"Since the great civil war the United States has made little progress in military ballooning. While other great nations notably the French, Germans and Russians have improved the system to a great degree of trustworthiness."*

Also, the first sentence of the article concludes the importance of Professor T.S.C. Lowe and his Corps' participation and contributions to the Union Army's Aeronautic program, *"Balloons have played an important part in war for more than a century."*

In conclusion, it may be said that even though the Union Army Balloon Corps was not utilized to its fullest extent, that it did indeed meet its goal with marked success. The American experiment led the countries of England, Germany, Russia, Austria and France down the path in developing their own balloon corps programs. It also set the tone for the future of aeronautics for decades to come, while spurring on the Wright Brothers and giving birth to their vision of manned flight.

Professor T.S.C. Lowe's Official Report

Professor T.S.C. Lowe,

Innovator, Pioneer, Inventor,

and

"Father of the United States Air Force"

Thaddeus Sobieski Constantine Lowe, founder of the United States Aeronautic Service, was born on August 20, 1832 in Jefferson Mills, New Hampshire. He was descended from 17th century New England Pilgrim stock on both his paternal and maternal side. Lowe's grandfather, Levi Lowe served as a

Professor T.S.C. Lowe's Official Report

militiaman and fought in the American Revolution and his father Clovis, participated in the War of 1812 and served as a drummer boy.

Young Thaddeus was brought up on the family farm and at the age of ten his mother Alpha passed away. Clovis Lowe re-married a woman by the name of Mary Randall and sent his young son to work on a neighboring farm with the specific purpose of learning a trade.
For the next four years, Thaddeus worked at his chores on the farm, but attended school only three months out of the year. However, he found the time to self educate himself through books which were loaned to him from his teacher's personal library.

Upon attaining his fourteenth birthday the youthful Lowe ventured out on his own and then traveled to Portland Maine to join his older brother in Boston, Massachusetts. He worked at various jobs and eventually was able to save some money from his employment as a shoe piece cutter.

Returning home and recuperating from a rather long illness, he attended a series of lectures given by professor Reginald Dinkelhoff featuring the newly developed theory of lighter-than-air-gases. He became very interested in the composition and properties of these gases, especially that of hydrogen.

He was offered an assistant's position by professor Dinkelhoff and continued to develop his knowledge not only of the properties of these lighter-than-air-gases, but of the construction and practical aspects of ballooning.

In 1857 Lowe built (with funds acquired from his own lectures and presentations) and piloted his first balloon in tethered flight over a farm in Hoboken, New Jersey. Later that year, his father joined him and together they built a large sized balloon which they named *Enterprise*. Additionally, the Lowe family had become pre-eminent as builders of balloons and more importantly

Professor T.S.C. Lowe's Official Report

established themselves as authorities in the use of hydrogen gas. They consistently continued to seek out new and improved methods regarding advanced balloon technology.

Lowe continued to improve his scientific methods and in 1859 began construction of a gargantuan balloon which would enable him to make a transatlantic flight via the high altitude currents. He continued to espouse the theory of a transatlantic crossing and raised additional funds for this endeavor.

On June 28, 1860, Lowe successfully piloted the balloon *Great Western* from Philadelphia to New Jersey. He attempted a transatlantic crossing on September 7, but the *GW* was damaged by high wind gusts.

He then made a series of test flights, which culminated on his most successful to date, when he took off on the morning of April 19, 1861 from Cincinnati Ohio and landed two days later in Unionville, North Carolina. Upon touchdown (and due to the fact that Fort Sumter had been fired upon) he was immediately arrested by southern authorities under the auspices of being a Union spy.

Lowe managed to convince them of his innocence and took a week of touring through the Confederate states under a letter of free passage to Cincinnati where he was to recover his balloons. While in Cincinnati he received word to see the Secretary of the Treasury and the War Department. After which he was referred to the President and then eventually General Scott

On the evening of June 11, 1861, Lowe met with President Abraham Lincoln and offered his services to the Union Army. He demonstrated the military application of ballooning as he sent a telegraph message to the White House via a 500 foot telegraph cable.

Professor T.S.C. Lowe's Official Report

The cable read, *"I HAVE THE PLEASURE OF SENDING YOU THIS FIRST DISPATCH EVER TELEGRAPHED FROM AN AERIAL STATION."*

Lowe and his balloon accompanied General Irvin McDowell at the first battle of Bull Run and provided impressive service to the Union Army. On July 25, 1861, President Lincoln impressed with the intelligence-gathering possibilities of manned balloon flights, sent a formal note to Lieutenant General Winfield Scott ordering him to form the United States Army Balloon Corps, with Lowe as Chief Aeronaut.

Over the nest two years, Lowe and his team of aeronauts would provide highly detailed information regarding Confederate Army troop movements and greatly improve upon the art of ballooning and its military applications.

Working under the auspices of the War Department, Lowe received the pay of a colonel, plus materials and labor. His first mission involved gathering information on Confederate troop deployment shortly after First Bull Run in mid-July 1861. During George B. McClellan's Peninsula Campaign in the summer and fall of the same year, Lowe conducted almost daily flights over the Virginia landscape, producing reports and photographs of the Confederate position.

Thanks to additional army appropriations, Lowe was able to expand and improve his fleet. He built five airships of various sizes and had with them newly designed generators that could produce hydrogen gas on the battlefield. The largest of his ships, the *Intrepid,* was 32,000 cubic feet in size and required 1,200 yards of silk ' He used it to conduct surveillance during and after Fredericksburg.

Although the Confederate Army lacked the resources to launch its own full-scale aeronautics program, Captain E. Porter Alexander oversaw several ascensions by Confederate aeronauts in 1861 and

Professor T.S.C. Lowe's Official Report

1862, who reported on Union troop deployment during the Peninsula and Seven Days campaigns. Balloons were often shot down behind enemy lines or, due to the unpredictable nature of balloon flights, were unable to return to camp in time to provide crucial information to the command. The last use of balloons by the Confederate army took place in 1863, after its largest balloon was swept away by a strong, high wind.

Under Lowe's superb leadership, innovative balloon designs and techniques became a common occurrence. Advanced technology such as portable hydrogen gas generators, an airborne military telegraph, the use of photographic reconnaissance and the first use of a balloon regarding seaborne operations, would pave the way for future generations.

In late 1862 and early 1863, Lowe experienced increasing opposition to the use of his Balloon Corps in the field. A disparaging third party report, which was refuted at length by Lowe, gave pause to the Union commanders for further use of balloons. Lowe tendered his resignation in May 1863 and due to inept management, the Balloon Corps had ceased to exist by August of the same year.

Lowe and his crew however had made more than 3,000 flights over enemy territory and supplied the Union Army with valuable information regarding the disposition of Confederate Forces.

Upon leaving federal service, Lowe made a new home in Norristown, Pennsylvania, not far from the Valley Forge encampment. He continued to experiment with hydrogen and improved upon and patented the water gas process, which created fuel by passing steam over hot coal. This process revolutionized the home heating and lighting industry and provided increased comfort to hundreds of thousands of families.

After a long illness, Thaddeus Constantine Sobieski Lowe died on January 29, 1913 at his daughter's home in Pasadena, California.

Professor T.S.C. Lowe's Official Report

The consummate innovator, pioneer and inventor, he was truly a man ahead of his time, continually striving for excellence in all that he attempted.

With an unrelenting determination to venture into uncharted territories, he will most assuredly be known as the man who passed on a legacy of integrity, self-sacrifice and excellence, and hold undisputed claim to the title, "FATHER OF THE UNITED STATES AIR FORCE."

Professor Thaddeus Lowe's Official Report - (Part I)

Professor T.S.C. Lowe's Official Report

Professor Thaddeus Lowe's Official Report (Part I)

O.R.--SERIES III--VOLUME III [S# 124] CORRESPONDENCE, ORDERS, REPORTS, AND RETURNS OF THE UNION AUTHORITIES FROM JANUARY 1, 1861 TO DECEMBER 31, 1863.--# 11

WASHINGTON, D.C., *June* 4, 1863.

Hon. E. M. STANTON,
Secretary of War:

SIR: I have the honor herewith to submit a report of all my operations from the commencement of the war to the present time, and in the hope that it may be found to contain valuable information and to establish beyond doubt the advantages of the aeronautic service to the Government.

I am, very respectfully, your most obedient servant,
T. S.C. LOWE,
Aeronaut.

[*Inclosure.*]

WASHINGTON, D. C., *May* 26, 1863.

Hon. E. M. STANTON,
Secretary of War:

Professor T.S.C. Lowe's Official Report

SIR: In accordance with your request I have the honor to submit the following report of my operations in the department of aeronautics, as connected with the military service of the Government:

Balloons have been employed for many years for the purposes of amusement or experiment, but they have never been constructed of durable materials, nor combined those qualities essential for frequent or long-continued observations, or for transportation from place to place, until the present war. The French were the first and only nation to make any use of this important means of securing information of the position and movements of an enemy, and even the imperfect apparatus they employed secured great advantages to them on two occasions. One of these was in June, 1794, when they were used for reconnoitering the position of the Austrians at the battle of Fleurus; the other was at the battle of Solferino, in 1859.

For nearly ten years my attention has been given to the subject of aeronautics, and I have made large expenditures in practical experiments to perfect and develop the system. Notwithstanding the fact that balloons were first invented in 1782, but little had been subsequently done to improve them. Various inventions of air ships had come into notice and proved to be impracticable, although the possibility of devising a means of navigating the air with safety was believed by many. Fully convinced of this myself, and that science and skill would produce the long-desired invention, I constructed a large balloon in 1859 for experiments, preliminary to an attempt to cross the Atlantic. This balloon when filled with gas would lift more than twenty tons in weight. The envelope alone weighed two and a quarter tons. Though treated as a visionary by the unthinking and by the timid, I received substantial aid and support from some of the most eminent scientific men of the country, and was thus encouraged to labor on in improving and perfecting every part of my apparatus, so that no reasonable ground of doubt should exist as to the ultimate success of the experiment.

In December, 1860, I presented the following memorial to the Smithsonian Institution, which I take the liberty of including in

Professor T.S.C. Lowe's Official Report

this report as an evidence of the favor with which my enterprise was looked upon by the distinguished men whose names are subscribed to it:

PHILADELPHIA, *December* --, 1860.

Prof. JOSEPH HENRY,
Secretary of the Smithsonian Institution, Washington, D. C.:

The undersigned, citizens of Philadelphia, have taken a deep interest in the attempt of Mr. T. S.C. Lowe to cross the Atlantic by aeronautic machinery, and have confidence that his extensive preparations to effect that object will add greatly to scientific knowledge. Mr. Lowe has individually spent much time and money in the enterprise, and in addition the citizens of Philadelphia have contributed several thousand dollars to further his efforts in demonstrating the feasibility of trans-Atlantic air navigation. With reliance upon Mr. Lowe and his plans, we cheerfully recommend him to the favorable consideration of the Smithsonian Institution, and trust such aid and advice will be furnished him by that distinguished body as may assist in the success of the attempt, in which we take a deep interest.

JNO. C. CRESSON.
WILLIAM HAMILTON.
W. H. HARRISON.
[AND THIRTEEN OTHERS.]

The Secretary of the Smithsonian Institution, to whom the memorial was referred, gave it a careful consideration, and although he did not recommend the appropriation of any of the

Professor T.S.C. Lowe's Official Report

funds of the Institution to assist me in constructing the balloon, stated the following as the result of his investigations:

It has been fully established by continuous observations collected at this Institution for ten years, from every part of the United States, that, as a general rule, all the meteorological phenomena advance from west to east, and that the higher clouds always move eastwardly. We are, therefore, from abundant observation, as well as from theoretical considerations, enabled to state with confidence that on a given day, whatever may be the direction of the wind at the surface of the earth, a balloon elevated sufficiently high would be carried easterly by the prevailing current in the upper or rather middle region of the atmosphere.

I do not hesitate, therefore, to say that, provided a balloon can be constructed of sufficient size and of sufficient impermeability to gas, in order that it may maintain a high elevation for a sufficient length of time, it would be wafted across the Atlantic. I would not, however, advise that the first experiment of this character be made across the ocean, but that the feasibility of the project should be thoroughly tested and experience accumulated by voyages over the interior of our continent.

In accordance with the last suggestion made by Professor Henry, and to remove all doubts from the minds of those who considered the risk of the ocean voyage too great, I made ascensions from points in the West, and had demonstrated the truth of my propositions, when the breaking out of the rebellion turned the thoughts of all loyal Americans to the state of the country. Feeling assured that I could render essential service to the Government in its time of need, and that my inventions would be appreciated by those who were in authority, I left Philadelphia on the 5th of June, 1861, for Washington, taking with me a new balloon with which I had made a voyage on the 20th of April of the same year from Cincinnati, Ohio, to the coast of South Carolina, from 4 a.m. to 1 p.m. of the same day, a distance of over 900 miles, in nine hours.

On arriving in Washington I immediately called on Professor

Professor T.S.C. Lowe's Official Report

Henry, who at once perceived the importance and value of my proposed operations. He had repeated interviews with the President of the United States, the Secretary of War (Mr. Cameron), and the officers of the Topographical Engineer Corps, and strongly urged the trial of experiments with my balloon to test its adaptation to the great work in which we were engaged. Discouragement and difficulty attended every effort, however, to secure attention; but finally, through the influence of Professor Henry, to whose disinterested and persevering support is in a great measure due the introduction of aeronautics into the military service of the United States, I was enabled to make preliminary experiments with the balloon I had brought to Washington.

The balloon was inflated from one of the gas mains in the Armory grounds, and repeated ascensions were made from that place, from the Smithsonian grounds, and from the front of the Executive Mansion. For the first time telegraphic communication was established between a balloon and the earth, and a message was sent to the President of the United States and others while at an elevation of a thousand feet.

For a detailed account of these experiments I have the honor to refer to the following report from Professor Henry, under whose supervision they were made:

SMITHSONIAN INSTITUTION,
June 21, 1861.

Hon. SIMON CAMERON:

DEAR SIR: In accordance with your request made to me orally on the morning of the 6th of June, I have examined the apparatus and witnessed the balloon experiments of Mr. Lowe, and have come to the following conclusions:

1st. The balloon prepared by Mr. Lowe, inflated with ordinary street gas, will retain its charge for several days.

2d. In an inflated condition it can be towed by a few men along an ordinary road, or over fields, in ordinarily calm weather, from the places where it is galled to another, twenty or more miles

Professor T.S.C. Lowe's Official Report

distant.

3d. It can be let up into the air by means of a rope in a calm day to a height sufficient to observe the country for twenty miles around and more, according to the degree of clearness of the atmosphere. The ascent may also be made at night and the camp lights of the enemy observed.

4th. From experiments made here for the first time it is conclusively proved that telegrams can be sent with ease and certainty between the balloon and the quarters of the commanding officer.

5th. I feel assured, although I have not witnessed the experiment, that when the surface wind is from the east, as it was for several days last week, an observer in the balloon can be made to float nearly to the enemy's camp (as it is now situated to the west of us), or even to float over it, and then return eastward by rising to a higher elevation. This assumption is based on the fact that the upper strata of wind' in this latitude is always flowing eastward. Mr. Lowe informs me, and I do not doubt his statement, that he will on any day which is favorable make an excursion of the kind above mentioned.

6th. From all the facts I have observed and the information I have gathered I am sure that important information may be obtained in regard to the topography of the country and to the position and movements of an enemy by means of the balloon now, and that Mr. Lowe is well qualified to render service in this way by the balloon now in his possession.

7th. The balloon which Mr. Lowe now has in Washington can only be inflated in a city where street gas is to be obtained. If an exploration is required at a point too distant for the transportation of the inflated balloon, an additional apparatus for the generation of hydrogen gas will be required. The necessity of generating the gas renders the use of the balloon more expensive, but this, where important results are required, is of comparatively small importance.

For these preliminary experiments, as you may recollect, a sum not to exceed $200 or $250 was to be appropriated, and in accordance with this Mr. Lowe has presented me with the inclosed

Professor T.S.C. Lowe's Official Report

statement of items, which I think are reasonable, since nothing is charged for labor and time of the aeronaut.

> I have the honor to remain, very respectfully, your obedient servant,
> **JOSEPH HENRY,**
> *Secretary Smithsonian Institution.*

On the evening of the 21st of June I received a telegram from Captain Whipple, of the Topographical Engineers, directing me to fill the balloon and to bring it, with the telegraphic apparatus, &c., to Arlington.

The gas could not be obtained from the Washington Gas Company until the following afternoon, when the balloon was inflated and taken across the Long Bridge to Arlington House, where, by order of Captain Whipple, it remained until the next morning at 4 o'clock, when I was ordered to take it to Falls Church. On arriving at the Alexandria and Loudoun Railroad I learned from the guards that there were no pickets out in the direction we were going. There being no other route by which the balloon could be towed, on account of the woods, and knowing the importance of observations from Falls Church, the balloon was let up by ropes to a sufficient height to ascertain that it was safe to proceed. We then advanced two miles farther, to Bailey's Cross-Roads, where I was informed by the residents that a rebel scouting party had just left, having seen the balloon, and supposing that a large force accompanied it. After stopping a few minutes we proceeded to Falls Church, where the balloon was kept in constant use for two days more, during which time General Tyler sent up an officer who sketched a fine map' of the surrounding country and observed the movements of the enemy. Captain Whipple and other officers also made several ascensions.

On the 26th of June I was informed by Captain Whipple that the Bureau of Topographical Engineers had concluded to adopt the balloon for military purposes, and desired me to furnish a full account of the method of operating the balloons in the field, and to make estimates for their construction, &c, The information I gave

Professor T.S.C. Lowe's Official Report

he noted down. The next day, upon calling on the captain to know what conclusion he had arrived at, I was informed that he had decided to give an order to Mr. Wise to construct a balloon, as his estimate was $100 or $200 less than mine, but that it was possible I might be employed to operate the balloon after it was made. To the latter part of his remarks I replied that I would not be willing to expose my life and reputation by using so delicate a machine, where the utmost care in construction was required, which should be made by a person in whom I had no confidence. I assured him that I had greater experience in this business than any other aeronaut, and that I would guarantee the success of the enterprise if intrusted entirely to my directions.

Feeling confident of the ultimate result, and not being willing to abandon my cherished plans for the benefit of the Government after so much expenditure of time and my own means, I instituted a series of experiments, on my own account, in the Smithsonian grounds, which brought together many officers and scientific men, who strongly recommended the adoption of my system of aeronautics. Among others who witnessed these experiments was Captain Whipple, who, finding that the balloon ordered from Mr. Wise had not arrived at the time promised, desired me to transport my balloon, then inflated, with the army which was moving toward Manassas. My operations at this time are described in the following communication addressed to Major Bache, of the Topographical Engineers, to which I would call particular attention:

WASHINGTON, *D.C., July* 29, 1861.

Major BACHE,
Bureau Topographical Engineers:

SIR: Having spent two months in Washington for the purpose of demonstrating the utility of balloon observations for war purposes, and thus far without any recompense, I feel it my duty before retiring from the seat of war to make a statement of what I have done and what might and can be accomplished, provided the

Professor T.S.C. Lowe's Official Report

Government would furnish the necessary means, which at most is very small compared with the results that can be attained.

In the first place, the balloon which I have been compelled to use (for want of a more suitable one) was intended for making free voyages, in which comparatively but little strength is required, and not for the purpose of ascension with ropes. On the 18th of June I inflated the balloon, and, with a telegraph apparatus attached, ascended with three persons and demonstrated the feasibility of communicating with the earth, which at times can be rendered very useful. This inflation lasted four days, although subjected to the pressure of several very heavy winds. Two days afterward the balloon was again inflated and transferred fourteen miles from the place of filling, and retained its charge for several days, during which time it was let up repeatedly, and on one occasion 1,000 feet with an officer, who sketched a map of roads and of the enemy's camps at Fairfax CourtHouse. Much greater results could have been obtained by making a free voyage at an altitude of a mile or two and returning in the upper current toward Washington. I then gave it another coat of varnish, which much increased its retentive power, and demonstrated the utility of the balloon for the purpose of reconnaissance to a number of gentlemen of this city on the Smithsonian grounds. After this I was suddenly required by Captain Whipple to fill my balloon and transport it into the interior of Virginia. Although this balloon was not intended for war purposes, and although I had cherished the hope of being directed to construct another, I concluded to do the best I could, and accordingly set about making the necessary preparations for the voyage; but when these were completed and I was ready to start, I was unable, on account of the absence of Captain Whipple, to procure the men and means for the inflation and transportation. Not being able to obtain assistance from Captain Whipple, who was then on duty, I concluded, on the advice of my friends, to inflate the balloon and procure men for its transportation on my own account, not doubting that my services would be properly appreciated; but to my disappointment I was informed by the director of the gas company that another balloon had arrived and was to be used instead of mine. On the receipt of this intelligence I

Professor T.S.C. Lowe's Official Report

removed my balloon from the inflating pipes, to give place to the other balloon, and ceased all further efforts until I was informed, on Sunday, that the competing balloon had proved a failure, and then being urged by several patriotic individuals, and hoping still to render some service to the army at Centerville or Manassas, I commenced on Sunday morning to make preparations for inflating and transporting my balloon, and on the evening of the same day started with it for Virginia. In this enterprise I was aided by the liberality of Colonel Small, who furnished me with twenty men from his command for the purpose. Unfortunately, when we arrived at Falls Church I was informed of the retreat of the army, and thinking it useless to attempt to go farther, I concluded to remain there, even after all the troops had passed by and in the midst of a drenching rain, with the hope that I might be of service in giving information as to the approach of the enemy; but as the pickets were withdrawn, I started again at 4.30 on Monday afternoon to return to Arlington, the rain continuing to fall in torrents, the wind against us, and arrived at Fort Corcoran at 8 o'clock the same evening with the balloon fully inflated, after having been transported against a wind of considerable force, through a distance in all of about twenty stories, the latter half of which was in a violent rain-storm. I remained with the balloon at Fort Corcoran until Wednesday morning, and then, taking advantage of the favorable weather, I ascended at 7.30 o'clock with an ascensional power of 500 pounds beyond the weight of the balloon itself. I obtained an altitude of about three and one-half miles and had a distinct view of the encampments of the enemy, and observed them in motion between Manassas Junction and Fairfax.

From the facts I have stated it must be evident to every one that the balloon can be rendered of essential service 'in the art of war, and that I have accomplished all I have undertaken without a single failure, with very imperfect means and with scarcely any aid from the Government.

Having thus given an account of what has been accomplished, I now proceed to furnish a statement of what might or can be done if proper facilities are afforded:

Professor T.S.C. Lowe's Official Report

First. It is very probable that balloons will be wanted for some time to come in the vicinity of Washington and Alexandria to watch the movements of the enemy and prevent a surprise. For this purpose the balloon now in my possession will answer very well until another can be procured. With it, almost every day or two, ascents can be made to a great altitude, affording an opportunity for several officers at the same time to observe, with good glasses, the position and movements of the enemy in perfect security, without risk of life or property.

Second. While the army is making preparation for another movement a lighter balloon, with portable apparatus, can be constructed in time to more with the troops, and be ready before and during an engagement to furnish the means of observations of the greatest importance.

Having made the necessary inquiries, I find that the required apparatus can be constructed by mechanics now in the Government employ in Washington; that the whole weight of material to inflate the balloon for several days' use will not exceed four tons, and can be carried in two or three wagons, and that the whole expense for inflating, aside from the apparatus, will not exceed $300, including transportation.

It will not be necessary to use this method of inflation, excepting at a distance from gas works too great to move an inflated balloon.

The same apparatus can also be used on the rivers, and ere long will probably be much wanted at Fortress Monroe, Norfolk, and Richmond, and many other places.

Should the Government conclude to adopt the above methods, and desire my services, I will give my plans in detail, and shall be pleased to carry them out. I can truly say that I have not, in my endeavor to introduce balloon observations into use in our Army, been governed by a desire for pecuniary gain, but I have been actuated by a wish to increase my reputation and advance the art to which I have devoted my life, by demonstrating its importance to the country in its present critical condition.

Hoping that if my services are further required, I may receive as early a notice as possible,

Professor T.S.C. Lowe's Official Report

I remain, very truly, your obedient servant,
T. S. C. LOWE,
Aeronaut.

The ascension of the 24th of July, alluded to in the foregoing letter, was made in consequence of a report being circulated that the enemy was marching in force on Washington, which caused much excitement. The result of my observations, published the next day, showed this report to be untrue and restored confidence.

In this voyage I started soon after sunrise, while the atmosphere was clear, and sailed directly over the country occupied by the enemy, as the lower current was blowing toward the west. Having seen what I desired, I rose to the upper current and commenced moving toward the east again, until over the Potomac, when I commenced to descend, thinking that the under current would take me back far enough to land near Arlington House. When within a mile of the earth our troops commenced firing at the balloon, supposing it to belong to the rebels. I descended near enough to hear the whistling of the bullets and the shouts of the soldiers to "show my colors." As I had, unfortunately, no national flag with me, and knowing that if I attempted to effect a landing there my balloon---and very likely myself--would be riddled, I concluded to sail on and to risk descending outside of our lines. This I accomplished, and landed on Mason's plantation, five miles and a half from Alexandria and two miles and a half outside of our pickets. A detailed account of my escape would be interesting, but it is sufficient to say that I was kindly assisted in returning by the Thirty-first Regiment New York Volunteers, and brought back the balloon, though somewhat damaged, owing to my having been obliged to land among trees. The balloon was generally supposed to be one of the enemy's, and the authorities in Washington were telegraphed from Arlington to this effect.

On the 29th of July I received the following dispatch from Captain Whipple:

Professor T.S.C. Lowe's Official Report

ARLINGTON, *July* 29, 1861.

T. S. C. LOWE:

If you will at once repair your balloon, and will superintend its transportation to this side of the Potomac, the United States will employ you temporarily as follows: The United States will pay for the gas used for the inflation, will furnish twenty men to manage the balloon, will pay you $30 per day for each day the balloon is in use for reconnaissance on the Virginia side of the Potomac. The balloon to be ready for use within twenty-four hours.

A. W. WHIPPLE,
Captain, Topographical Engineers.

In answer to this I informed Captain Whipple that I could not enter upon such an arrangement, but that if the Government would direct me to construct a balloon such as I deemed suitable for military purposes I would only charge $10 a day for my services, instead of $30, and would guarantee entire success. I also stated the cost of the new apparatus and the time required for its construction.

I, however, repaired the balloon, as desired by Captain Whipple, but while transporting it with inexperienced men a distance of ten miles over a rough road, where there were many obstructions, we were overtaken by a heavy storm and I was obliged to discharge the gas. In relation to this occurrence I beg leave to refer to the following letter from Professor Henry:

SMITHSONIAN INSTITUTION, *August* 2, 1861.

Capt. A. W. WHIPPLE,
U. S. Army:

Professor T.S.C. Lowe's Official Report

DEAR SIR: I regret very much to learn from Mr. Lowe that you think of giving up the balloon operations, and I write to express the hope that you will make further attempts. A single successful observation will fully repay all that you have yet expended.

The experiment of Wednesday was rendered abortive by the accidental occurrence of a thunder-storm which could not be foreseen. At this season of the year thunder-storms occur generally in the after part of the day or night, and I would therefore advise that the balloon be filled immediately after the clearing off of the sky, and then used as soon as possible after daylight the next day.

Mr. Lowe came to this city with the implied understanding that, if the experiments he exhibited to me were successful, he would be employed. He has labored under great disadvantages, and has been obliged to do all that he has done, after the first experiment, without money. From the first he has said that the balloon he now has was not sufficiently strong to bear the pressure of a hard wind, although it might be used with success in favorable situations and in perfectly calm weather. I hope that you will not yet give up the experiments, and that you will be enabled with even this balloon to do enough to prove the importance of this method of observation, and to warrant the construction of a balloon better adapted to the purpose.

I remain, very truly, your obedient servant,
JOSEPH HENRY.

Up to this time I had used my own machinery, and had a party of persons constantly employed at my own expense to assist in the management of the balloon and to keep it in order.

On the 2d of August I called on Maj. Hartman Bache and gave him a detailed account of what I had accomplished, also setting forth the advantages of using balloons, provided proper facilities were afforded. Upon this Major Bache gave me a letter to Captain Whipple, authorizing him to direct me to construct such a balloon as I desired; upon the receipt of which the latter gave me the following order and instructions:

Professor T.S.C. Lowe's Official Report

HEADQUARTERS DEPARTMENT OF NORTHEASTERN VIRGINIA,
Arlington, August 2, 1861.

Mr. T. S. C. LOWE,
Aeronaut:

SIR: You are hereby employed to construct a balloon for military purposes capable of containing at least 25,000 cubic feet of gas, to be made of the best India silk, not inferior to the sample which is divided between us, you retaining a part, with best linen network, and three guys of manilla cordage from 1,200 to 1,500 feet in length. The materials you will purchase immediately, the best the markets afford and at prices not exceeding ordinary rates; and the bills you will forward to me through Maj. Hartman Bache, chief of the Corps of Topographical Engineers. When these materials shall have been collected at Philadelphia, where the balloon is to be constructed, you will report to me, that I may send an officer of the corps to inspect them. You need not, however, wait for the inspecting officer, but go on rapidly with the work, with the understanding that it may be suspended, provided that upon examination the materials or work prove unsatisfactory.

Your compensation from the day of collecting the materials and during the time of making the balloon shall be $5 per day, provided that a reasonable time be allowed for the collection and ten days for making. From and after the day that the balloon shall be ready for inflation at Washington, D.C., your compensation will be $10 per day as long as the Government may require your services.

Inclosed herewith is an order authorizing the purchase of materials necessary for the operation with which you are charged.

Very respectfully,
A. W. WHIPPLE,
Captain, Topographical Engineers.

Professor T.S.C. Lowe's Official Report

[Inclosure.]

HEADQUARTERS DEPARTMENT OF NORTHEASTERN VIRGINIA.

Mr. T. S. C. Lowe, aeronaut, is hereby authorized to purchase 1,200 yards of best India silk and sufficient linen thread, cordage, &c., for the construction of a balloon, and all reasonable bills for the same, when presented to me through the Bureau of Topographical Engineers, will be paid.

A. W. WHIPPLE,
Captain, Topographical Engineers.

From this time until the 28th of August was consumed in the construction of the first substantial war balloon ever built.

The main obstacle to the successful use of balloons still had to be overcome, namely, a portable apparatus for generating the gas in the field. I had already devised a plan for this purpose, but, as I could not then obtain an order to construct the apparatus, I was obliged to inflate the balloon as formerly in Washington, and to confine its operations to that locality. At this time I received the following orders from Major Woodruff and Captain Whipple:

WASHINGTON, *D.C., August* 28, 1861.

Mr. LOWE,
Balloonist, Washington, D. C.:

SIR: Get the silk balloon in readiness for inflation immediately. A detail of thirty men will repair to the Columbian

Professor T.S.C. Lowe's Official Report

Armory to aid you in the inflation and transportation of the balloon.

Respectfully, yours,
I. C. WOODRUFF,
Major, Topographical Engineers.

Inclosed is an order for gas.
I. C. W.

FORT CORCORAN, *August* 29, 1861.

Prof. T. S. C. LOWE:

The general desires you to be here at 3 a.m. to-morrow morning to make an ascension before daybreak to examine camp-fires, and ascend again as soon as it may be light enough to watch for movements of any bodies of men. Should I not be present please write the observations and send them to me by express at Arlington.

A. W. WHIPPLE,
Captain, Topographical Engineers.

These orders were complied with, and during my observations I discovered the enemy for the first time building earth-works on Munson's Hill and Clark's Hill, and also saw their movements along the entire line. In the afternoon I moved the balloon to Ball's CrossRoads and there took several observations, during which the enemy opened their batteries on the balloon and several shots passed by it and struck the ground beyond. These shots were the nearest to the U.S. capital that had been fired by the

Professor T.S.C. Lowe's Official Report

enemy, or have yet been, during the war.

From this time the balloon was kept in constant use and daily reports made to the commanding officers. I regret that I kept copies of but few of these, as at the time I did not consider that they would be required. Confidence in this new means of observation soon began to be manifested, and many officers made ascensions, among whom were Generals McDowell, Porter, and Martindale. On the 7th of September Major-General McClellan ascended and made an examination of the enemy's works on Munson's Hill and other points, a view of which it was impossible to obtain by any other means.

From this time to the 27th of September many alarms were given, and the troops called out in line of battle, and in every instance after an examination had been made by means of the balloon the troops were sent back to their quarters and allowed to rest without danger of being surprised.

Having only one balloon, I was necessarily compelled to lose some time to go to Washington for gas, which I invariably did, however, at night.

The following papers will indicate my operations to the 27th of September:

HEADQUARTERS PORTER'S DIVISION,
Fort Corcoran, Va., September 7, 1861.

Professor LOWE:

SIR: General Porter desires you to make a reconnaissance in your balloon as early as possible to-morrow morning. Be kind enough to send the result of your observations to General Porter, whether you discover anything of interest or not.

Very respectfully,
J. F. McQUESTEN.

BALLOON HEADQUARTERS, *September* 8, 1861.

Professor T.S.C. Lowe's Official Report

Brigadier-General PORTER,
Commanding Division, Fort Corcoran :

DEAR SIR: According to your request I made two reconnaissances with the balloon this morning. The first a little after 4 o'clock. At that time no lights were visible in the west. At 5 o'clock one light to the right of Munson's Hill and one at Taylor's Corners appeared, which were all that could be seen. I ascended again at 6 o'clock and had a clear view of the works on Munson's Hill, also Upton's, but observed nothing unusual, the strong wind preventing me from attaining an altitude to observe with distinctness anything beyond these points. I will ascend again during the day and report to you.

Your obedient servant,
T. S. C. LOWE,
Aeronaut.

HEADQUARTERS PORTER'S DIVISION,
Fort Corcoran, Va., September 9, 1861.

Professor LOWE,
Fort Corcoran:

PROFESSOR: General McClellan desires you to transfer your balloon to the Chain Bridge early to-morrow to take observations. I have informed him you will inflate as early as practicable and move up to Chain Bridge. I am desirous to see you prosper, and I think you are now on the road. I have recommended an increase of two balloons and movable inflating apparatus, and as soon as the utility of the science is made apparent (which will depend on your energy) I have no doubt of success. Strike now while the iron is hot. I suggested your balloon should be sent up to Chain Bridge or its vicinity, and I doubt not General McClellan will be there, or others, who will work for you if they are satisfied

Professor T.S.C. Lowe's Official Report

of its utility. General Smith is in command, and I promise a good reception for you.

If I can aid you in any manner, don't hesitate to call. I will be pleased to see you before you go over in the morning, and the result of your morning observation, which I beg of you to take.

> I am, sir, with great respect, your obedient servant,
> **F. J. PORTER.**

BALLOON HEADQUARTERS, *September* 9--2.30 *p.m.*

General PORTER:

I have just concluded another observation with the balloon and had a distinct view of Falls Church.

In answer to your inquiry, I can say that there is no appearance of the enemy in or about Falls Church other than has been reported before. Munson's Hill and other places remain the same.

> Very respectfully, your obedient servant,
> **T. S. C. LOWE.**

FORT CORCORAN, *September* 11, 1861.

Professor LOWE:

I have nothing special. As your balloon is near Chain Bridge, I suggest you ask General Smith if he has anything. I presume if you can rise in the morning he would like it. You are of value now.

> **F. J. PORTER.**

Professor T.S.C. Lowe's Official Report

ARLINGTON, *VA., September* 16, 1861.

Brig. Gen. F. J. PORTER,
Commanding Division at Fort Corcoran:

SIR: In accordance with your request I herewith send a statement of what I should advise and deem necessary in addition to the means now at hand for the purpose of facilitating and making more frequent reconnaissances with balloons, and from various points at the same time, also for the purpose of being ready to accompany the army whenever a movement is made.

First. An addition of two balloons would be required, with capacities as follows: One of 30,000 cubic feet and one of 20,000, built of the best India silk and linen cordage, with all my late improvements and appliances. The cost of these air vessels complete will be, for the largest, $1,500; the smallest, $1,200.

Secondly. A portable inflating apparatus would be required for the purpose of inflating a balloon at any point where common gas cannot be obtained, and also for the purpose of replenishing the balloons when the gas is partially expended. This would save the expense of an entire reinflation, and also keep the balloon ready for observation at all times; besides, the hydrogen being more buoyant than coal gas, a greater altitude can be obtained.

The whole cost of this apparatus ought not to exceed $500, and can be built by ship carpenters and coppersmiths now in the employ of the Government at Washington. The time required for getting up these balloons and apparatus will be about two weeks, perhaps less, should the weather prove fine while coating the material.

By being supplied with the above additional equipments I feel confident in being able to keep the Government constantly informed of the movements and position of the enemy, as well as the topography of the country. Wherever occasion requires, the balloons can also be used for letting up various colored signal lights at night, which can be made to burn for a long time, and

Professor T.S.C. Lowe's Official Report

consequently will be seen with more certainty than by any other means.

Very respectfully, yours,
T. S. C. LOWE.

BALLOON HEADQUARTERS, *September* 20, 1861.

Brig. Gen. F. J. PORTER,
Commanding Division, Fort Corcoran:

DEAR SIR: I have just taken an observation from an altitude of 1,000 feet, and find the atmosphere uncommonly clear in the west. I shall move to the place where you first ascended, and would be pleased if you can come and go up with me. We may be able to discover something of interest.

Very respectfully, your obedient servant,
T. S. C. LOWE.

BALLOON HEADQUARTERS, *September* 22, 1861.

Brigadier-General PORTER,
Commanding Division, Fort Corcoran:

During my observations this evening I noticed a pretty heavy picket force on Upton's Hill and several camp smokes at Taylor's Corners. On the west slope of Munson's Hill there appeared to be a full regiment with a set of colors, their bayonets glistening in the sun as if on parade. I could see nothing of the horses you spoke of, but as soon as I can get the balloon inflated again I will go nearer and examine the woods.

Professor T.S.C. Lowe's Official Report

Very respectfully, yours,
T. S. C. LOWE.

CAMP ADVANCE, *September* 23, 1861.

General F. J. PORTER:

 At about 8.30 to-morrow morning I wish to fire from here at Falls Church. Will you please send the balloon up from Fort Corcoran and have note taken of the position reached by the shell, and telegraph each observation at once.

W. F. SMITH.

HEADQUARTERS PORTER'S DIVISION,
Fort Corcoran, Va., September 24, 1861.

Professor LOWE:

 SIR: By direction of General Porter I herewith inclose a telegram from General Smith. It explains itself. Two mounted orderlies will be sent you so that you can, with the assistance of your officer, report and send to these headquarters. During the time of fire it is very important to know how much the shot or shell fall short, if any at all.

Very respectfully, yours,
J. F. McQUESTEN,
Lieutenant and Aide-de-Camp.

[Inclosure.]

SEPTEMBER 24, 1861.

Professor T.S.C. Lowe's Official Report

General F. J. PORTER:

If we fire to the right of Falls Church, let a white flag be raised in the balloon; if to the left, let it be lowered; if over, let it be shown stationary; if under, let it be waved occasionally.

W. F. SMITH.

HEADQUARTERS OF BALLOON,
Arlington, September 24, 1861.

Brig. Gen. F. J. PORTER,
Commanding Division, Fort Corcoran:

SIR: This evening I took the balloon out near Ball's Cross-Roads and remained up nearly two hours. I had a distinct view of the works on a hill about one mile and a half beyond Munson's Hill. There seems to be heavy guns mounted and a pretty heavy force near by. Several tents were visible about there and a number of bodies of men on parade.

To the left of a high bluff, and about ten miles distant to the left, or nearly in a line with Bailey's Cross-Roads, there appeared to be a long line of smoke, as.if there were several camps. The smokes of the enemy's pickets are quite numerous, and a large body of men were on Upton's Hill, and also what appeared to be a field piece.

The whole distance from Chain Bridge to Falls Church is shown plainly from my new point of observation, and I think a shell could not be fired without seeing where it strikes.

Should it be convenient for you to come and go up in the morning the first thing, I think you will gain some valuable information.

Professor T.S.C. Lowe's Official Report

Very respectfully, your obedient servant,
T. S. C. LOWE.

CHAIN BRIDGE, *September* 24, 1861.

General PORTER:

I am going to Lewinsville to-morrow (Wednesday). Will you let Professor Lowe go up at 11, or little before, to watch the road from Falls Church and round to Lewinsville? Can't practice at fort to-morrow.

W. F. SMITH.

CHAIN BRIDGE, *September* 25, 1861.

Professor LOWE:

General Smith desires you to go up in the balloon this morning to observe the movements of troops, although we will not fire from the fort. The general is going out with the command, and firing will only be done in case the enemy is met.

C. MUNDEE,
Assistant Adjutant-General.

SEPTEMBER 25, 1861.

PROFESSOR: Look out for a battle at Lewinsville, and movements between us and that point.

F. J. PORTER.

Send me word of anything important.

Professor T.S.C. Lowe's Official Report

SEPTEMBER 25, 1861.

PROFESSOR: I am anxious about the movement from Chain Bridge. The enemy has moved north and has all his force between General Smith and Lewinsville, evidently to intercept his return.

I wish to get as much information of his movements, or what is transpiring, possible before sundown. I expect the return of the enemy, and if much dust be visible wish to know it, that I may send out a force.

F. J. PORTER.

If you can get up against this wind, will be glad. An important move is on foot.

HEADQUARTERS OF BALLOON, *September* 25, 1861.

Brigadier-General PORTER,
Commanding Division, Fort Corcoran:

SIR: Soon after you departed I heard the report of three guns toward Chain Bridge. I ascended and remained up until 12 o'clock, during which time no more guns were fired. About three miles in advance of Chain Bridge I could distinguish the glistening of bayonets and quite a large body of men in motion, but as they were going from the bridge I concluded they were General Smith's forces.

The parade at the Seminary made a grand display, while on Munson's Hill quite a large crowd were gathered. After descending I heard two more guns in the direction of the Chain Bridge, but the wind has arisen and prevents me from taking any observations at present. I am confident that there is no great movement on the part of the enemy, or I should have seen

Professor T.S.C. Lowe's Official Report

something of it, although the distance and heavy smoke are great obstacles to-day in that direction.

Very respectfully, your obedient servant,
T. S. C. LOWE.

QUARTERMASTER-GENERAL'S OFFICE,
Washington City, September 25, 1861.

Prof. T. S. C. LOWE:
(In care Maj. S. Van Vliet, *senior quartermaster, Army of the Potomac, Washington.*)

SIR: Upon the recommendation of Major-General McClellan the Secretary of War has directed that four additional balloons be at once constructed under your direction, together with such inflating apparatus as may be necessary for them and the one now in use. It is desirable that they be completed with the least possible delay.

Very respectfully, your obedient servant,
M. C. MEIGS,
Quartermaster-General.

On the 30th of September the balloon was taken to Upton's Hill and used constantly, General McDowell, the Count de Paris, and other officers ascending with me and gaining much valuable information greatly needed at the time, as there was no other means of learning the position and movements of the enemy, and where an attack was expected. I received many complimentary remarks during the day from the officers, who were satisfied of the value of the balloon for reconnaissances.

From the 1st to the 12th of October the balloon was left in charge of an assistant while I was engaged in the construction of the balloons and gas generators ordered by the Secretary of War.

Professor T.S.C. Lowe's Official Report

QUARTERMASTER-GENERAL'S OFFICE,
Washington City, October 1, 1861.

Lieut. Col. G. H. CROSMAN,
Deputy Quartermaster-General, Philadelphia, Pa.:

COLONEL: The Secretary of War having authorized Professor Lowe to construct four balloons for military purposes, you will pay for them, and such bills as may be made by him in their construction, the whole amount to be paid being about the sum he names as their cost, viz, for the two largest $1,500 each, and for the smallest $1,200 each.

Very respectfully, your obedient servant,
By order:
E. SIBLEY,
Brevet Colonel, U. S. Army, Deputy Quartermaster-General.

GENERAL MCCLELLAN'S HEADQUARTERS,
Washington, October 12, 1861.

Professor LOWE:

General McClellan directs that you report yourself to General Smith at Johnson's Hill. Be there sure to-morrow, Sunday night.

A. V. COLBURN.

In accordance with the above order I inflated the balloon the same evening and started at 9 p.m. Our progress was slow, the night being very dark, and we were constantly apprehensive of running the balloon against trees or other obstacles. After passing through Washington and Georgetown, crossing numerous flag ropes and telegraph wires stretched across the streets, we reached

Professor T.S.C. Lowe's Official Report

the road to the Chain Bridge. This was lined with trees and we were compelled to go across the fields, as the wind was too high to tow the balloon when elevated, and it soon became cloudy and so dark that it was with the utmost difficulty we advanced. At several points trees had to be felled to allow a passage for the balloon. We arrived at the Chain Bridge about 3 o'clock the next (Sunday) morning, and found it filled with artillery and cavalry going to Virginia. In order to take the balloon over my men were obliged to mount the trestle-work and walk upon the stringers, only eighteen inches wide and nearly 100 feet above the bed of the river. Thus, with the balloon above their heads, myself in the car directing the management of the ropes, the men getting on and off the trestle-work, with a column of artillery moving below, and 100 feet still lower, the deep and strong current rushing over the rocks, while the sky was dark above, the scene was novel, exciting, and not a little dangerous. At daybreak we arrived near Lewinsville, nearly exhausted by the excessive fatigue of the trip. Here a strong wind sprung up suddenly and I was obliged to lash the balloon with strong ropes to stumps in a field. In a few minutes the wind increased to a terrific gale, which continued for an hour, tearing up trees by the roots close to where the balloon was anchored. When the storm reached its height the cordage gave way and the balloon escaped. It ascended to a great height, and in less than an hour landed to the eastward on the coast of Delaware, a distance of about 100 miles, where I afterward obtained it. This gale proved the great strength of the balloon silk, and that the cordage was insufficient in comparison, although it was capable of bearing a strain of ten tons. I immediately ordered all the rest of the cordage used for my balloons to be made strong enough to resist a strain of twenty-five tons, which has proved sufficient to resist any gale thus far.

    From this time to the 10th of November I was occupied in superintending the construction of balloons and gas generators. From the latter date to the end of the year the following reports and communications (to which I would call attention) embrace the principal operations in which I was engaged.

Professor T.S.C. Lowe's Official Report

BALLOON EXPEDITION ON BOARD
U. S. STEAM TUG CŒUR DE LION,
Mouth of Mattawoman Creek, Sunday Evening, November 10, 1861.

Major-General HOOKER:

SIR: In obedience to orders of Major-General McClellan I have come to this place for the purpose of making an aeronautic observation of the forces of the enemy. The balloon will be inflated immediately, so as to be ready for use early to-morrow morning.

Will you have the kindness to detail an officer to confer with me, so that I may make such dispositions and arrangements as will best enable me to accomplish the object for which I have been sent.

Very respectfully, your obedient servant,
T. S. C. LOWE,
Chief Aeronaut, U. S. Army.

NAVY-YARD,
Washington, D. C., November 12, 1861.

Lieutenant-Colonel COLBURN:

DEAR SIR: I have the pleasure of reporting the complete success of the first balloon expedition by water ever attempted. I left the navy-yard early Sunday morning, the 10th instant, with a lighter (formerly the G. W. P. Custis) towed out by the steamer Coeur de Lion, having on board competent assistant aeronauts, together with my new gas generating apparatus, which, though used for the first time, worked admirably. We located at the mouth of Mattawoman Creek, about three miles from the opposite or

Professor T.S.C. Lowe's Official Report

Virginia shore. Yesterday I proceeded to make observations, accompanied in my ascensions by General Sickles and others. We had a fine view of the enemy's camp-fires during the evening, and saw the rebels constructing new batteries at Freestone Point. I was under the necessity of returning for some necessary articles this morning, and will go back immediately to continue in person the reconnaissances.

After making all necessary arrangements below, and leaving a competent aeronaut and assistants in charge, I shall return and place the other balloons wherever the general desires them. I have now a competent aeronaut for each of the new balloons, and in the course of a few days they can all be in active operation. I will call and see you on my return.

Your obedient servant,
T. S. C. LOWE,
Aeronaut.

HEADQUARTERS OF THE ARMY,

November 16, 1861.

Professor LOWE:

General McClellan desires me to say that he desires to have the first balloon kept ready to be sent to Port Royal; the second one he desires to have sent to Brigadier-General Stone, at Poolesville, as soon as it is ready.

I am, sir, very respectfully, your obedient servant,
A. V. COLBURN,
Assistant Adjutant-General.

Professor T.S.C. Lowe's Official Report

HEADQUARTERS OF THE ARMY,
Washington, November 16, 1861.

Professor LOWE:

General McClellan desires that you have a balloon ready to be taken to Port Royal by the first opportunity. It is impossible to tell exactly when it can be sent, but I will try to give you three or four days' notice.

Very respectfully,
A. V. COLBURN,
Assistant Adjutant-General.

A report was circulated that the enemy were advancing their forces, and I was ordered to make a reconnaissance, of which the following was the result:

NATIONAL HOTEL,
Washington, November 21, 1861.

Lieut. Col. A. V. COLBURN:

DEAR SIR: Yesterday I inflated one of the balloons, the Intrepid, and moved it to Minor's Hill. It being too late for taking observations last night, I ascended at daybreak this morning, and remained up until 8 o'clock, which was sufficient to ascertain that the enemy is not in force this side of Centerville. Judging from our own camp-fires and smokes, I should say there may be three or four regiments at Fairfax Court-House; twice that number at Centerville and more at Manassas, but nothing like the amount of smokes from our own camps in General Porter's division.

Their line of picket smokes near the line of the Leesburg turnpike was quite regular, and occasionally pickets could be seen in the roads and clearings, but owing to the haziness of the atmosphere no moving bodies of troops or their tents were visible.

Professor T.S.C. Lowe's Official Report

The balloon for the South is all ready. Can you tell me from what place I shall ship the materials for making gas? If from here I must have them sent from Philadelphia to this city, that they may be ready.

I intend going down the river to-morrow to reinflate the balloon at Budd's Ferry. By that time the apparatus for Poolesville will be ready, and I will station one there also.

Very respectfully, your obedient servant,
T. S. C. LOWE.

HEADQUARTERS OF THE ARMY,
November 22, 1861.

Prof. T. S. C. LOWE:

General McClellan desires that you send a balloon to Fort Monroe this evening or at latest by to-morrow evening's boat to go to Port Royal. The transports will leave Fort Monroe day after to-morrow.

A. V. COLBURN,
Assistant Adjutant-General.

If Captain Craven can spare the Coeur de Lion, and Captain Dahlgren also, the Department agrees to allow her to take Professor Lowe to Old Point.

G. V. FOX,
Assistant Secretary.

WASHINGTON, *November* 23, 1861.

Professor T.S.C. Lowe's Official Report

Major-General HOOKER,
Budd's Ferry, Md.:

    I start for Fortress Monroe to-morrow afternoon. Will take the balloon-boat down with me. Please inform me at what point I can anchor where it will be safe, and will be of the most service to you.

T. S. C. LOWE,
Chief Aeronaut, U. S. Army.

BUDD'S FERRY, *November* 24, 1861.

Professor LOWE:

    The safest and most convenient place for anchoring your steamer will be about one mile below your former anchorage. The balloon is now near the Posey house, and it is from that point I desire to make the next ascension if agreeable to yourself.

JOSEPH HOOKER,
Brigadier-General.

OLD POINT, *VA., November* 27, 1861.

Brig. Gen. T. W. SHERMAN,
Commanding Forces at Port Royal, S. C.:

    SIR: By direction of General McClellan I send to your command a balloon and aeronautic apparatus in charge of Mr. J. B. Starkweather, aeronaut, who will report to you for service. For the purpose of aiding in these operations Mr. Starkweather will require thirty men and a good officer. Should it be necessary to take observations at various points, there will be required two

Professor T.S.C. Lowe's Official Report

ordinary army wagons to convey the gas generators and materials. Anything further that will be required will be made known by the aeronaut.

Very respectfully, your obedient servant,
T. S. C. LOWE,
Chief Aeronaut, U. S. Army.

HALL'S HILL, *November* 30, 1861.

Professor LOWE:

Promise of a fair day to-morrow. Your balloon is wanted, and it is of the highest importance that it should be here to take advantage of the first calm. Can it be here early in the morning? I will send in men now if you will send it.

F. J. PORTER.

WASHINGTON, *November* 30, 1861--11.45 *p.m.*

Brig. Gen. F. J. PORTER,
Hall's Hill, Va.:

Please send in the men and I will do my best to get the balloon there. The inflating apparatus, as fast as finished so far, has been ordered to other points, or I would make the gas on the ground; but for this time I must tow it, as soon as the men get here.

T. S. C. LOWE,
Aeronaut.

Professor T.S.C. Lowe's Official Report

NOVEMBER 30, 1861.

General HOOKER:

General McClellan desires me to get a map of the enemy's position opposite your command. Can you accommodate me by sending up a draughtsman, and forwarding the result to the general?
This fine weather will not last long. Please have the aeronaut improve every opportunity.

T. S. C. LOWE,
Chief Aeronaut.

WASHINGTON, *December* 1, 1861.

WILLIAM PAULLIN,
In Charge of Balloon, Budd's Ferry:

Do not reinflate the balloon until it has another coat of varnish, unless it is perfectly tight. I will send you an assistant with all the necessary articles to-morrow. Improve every calm hour from daybreak until dark. Examine the shore opposite Mattawoman Creek, and keep me constantly informed.

T. S. C. LOWE.

WASHINGTON, *D.C., December* 3, 1861.

Lieut. Col. A. V. COLBURN,
Assistant Adjutant-General:

DEAR SIR: I have the honor to communicate to you the disposition thus far of the new balloons in my charge. The balloon

Professor T.S.C. Lowe's Official Report

Constitution is at Budd's Ferry--General Hooker's division. The Washington, with gas generating apparatus and materials, is *en route* for Port Royal, S.C. The Intrepid, of larger dimensions, is at General Porter's division, Hall's Hill. The Union, same size, is intended for Poolesville, and is now ready, but has been delayed at the navy-yard for work on gas-generating apparatus that was promised me three weeks ago. It was supposed to be a matter of economy to have this apparatus constructed at the navy-yard. This season of the year is not the most propitious for continued reconnaissances, but when all the work now under my supervision is completed, no favorable opportunity for observations, night or day, will be allowed to pass unimproved.

    I have thus far exercised, and in the future shall continue to exercise, the most untiring diligence in the prosecution of the important labors intrusted to me; but, in my judgment, the interests of this branch of service require the immediate construction of two small balloons, for the following, among other reasons, which I herewith respectfully commend to your favorable consideration: When General McClellan recommended, and the Secretary of War ordered, the addition of four balloons, the possibility or probability of using either of them at the South was not considered; therefore, as the ample supply of coal gas at Washington justified me in doing, I made two of them of larger dimensions, so that being filled with coal gas they would economically accomplish the equivalent of the work expected from a smaller envelope filled with hydrogen, notwithstanding the difference in levity of the two gases. These two small hydrogen balloons, as compared with the larger ones, will be particularly serviceable at the present time, as they will require one wagon leas each for moving generators, while the diminished amount of material required will also tax our transportation facilities to a much less extent.

    Lastly, the most important advantage gained will be that a light balloon, of small dimensions, well filled with hydrogen, presents so much less surface to the wind, and can consequently be used in the heavier weather. These qualities are embraced in the balloons Washington and Constitution.

    Hoping the general will allow me to construct the two small

Professor T.S.C. Lowe's Official Report

balloons, while the larger ones are held in reserve as future contingencies may determine,

> I remain, dear sir, very respectfully, your obedient servant,
> **T. S. C. LOWE.**

WASHINGTON, *D.C., December* 10, 1861.

Lieut. Col. A. V. COLBURN,
Assistant Adjutant-General:

DEAR SIR: One of my assistants arrived this morning from General Hooker's headquarters and reports that the balloon has been constantly used for the past week making observations of the enemy's movements and position. A large number of ascensions have been made, the aeronaut being accompanied by Colonel Cowdin, Colonel Small, and others. Colonel Small while up with the balloon made a very fine map of the enemy's works and surrounding country, a copy of which is being prepared, and will be forwarded to headquarters.

> Very respectfully, your obedient servant,
> **T. S. C. LOWE,**
> *Chief Aeronaut.*

WASHINGTON, *D. C., December* 16, 1861.

Lieut. Col. A. V. COLBURN,
Assistant Adjutant-General:

DEAR SIR: I returned yesterday from Poolesville, after stationing a balloon and necessary inflating apparatus with

Professor T.S.C. Lowe's Official Report

General Stone's division. This is the third of the new inflating apparatus which has been sent out, and three more are now ready to go as soon as the other two balloons are finished. I commenced inflation at Edwards Ferry on Friday at 4 p.m., and in three hours generated gas sufficient to lift 1,200 pounds.

On Saturday morning I ascended quite early and took an observation of the enemy's country. Very few troops were visible, and these were scattered both up and down the river. We could see into nearly every street of Leesburg, but scarcely any troops were visible. The main body appears to be between Leesburg and Centerville--I should judge fifteen or twenty miles below the former--as camps and heavy smokes were quite visible in that direction.

Later in the day I ascended again, and a number of their tents which were visible in the morning inside of their earth-works between Edwards Ferry and Leesburg were taken down, and teams were observed moving toward the village of Leesburg.

In the afternoon I was accompanied in my ascension by General Stone, who added several points to his map. The balloon still remains inflated, and will be ready for use at all times, in charge of a competent assistant aeronaut. The balloon now located at Budd's Ferry has been inflated over two weeks without any replenishing.

The communication of W. G. Fullerton, of December 2, in reference to photographic pictures taken from the balloon which was referred to me, has been examined, and I would say that the author advances no new ideas. As soon as other matters connected with the balloons are accomplished I shall give the photographic matter a thorough and practical test.

Very respectfully, your obedient servant,
T. S. C. LOWE,
Aeronaut.

During the months of January and February balloons were kept in constant use at Budd's Ferry, Md., under the orders of General Hooker; at Poolesville, Md., General Stone's command; at

Professor T.S.C. Lowe's Official Report

Port Royal, S.C., General Sherman's command, and there was one also sent to Cairo, Ill. The one last mentioned was used by Commodore Foote at the attack on Island No. 10. During the bombardment an officer of the Navy ascended and discovered that our shot and shell went beyond the enemy, and by altering the range our forces were soon able to compel the enemy to evacuate.

Up to the 1st of March I was principally occupied in visiting the different balloon stations and keeping everything in order. As the reports were made directly to the officers in command of the posts where the balloons were stationed, I can only furnish the following communications:

POOLESVILLE, *January* 20, 1862.

Professor LOWE,
National Hotel, Washington:

Please send up the small balloon immediately. The large one has suffered in its varnish from the excessively bad weather.

C. P. STONE,
Brigadier-General of Volunteers.

POOLESVILLE, *January* 25, 1862.

Professor LOWE,
National Hotel, Washington:

The balloon Intrepid got an inch of ice on it last night and is reported much injured. Hurry up the smaller one.

C. P. STONE,
Brigadier-General.

Professor T.S.C. Lowe's Official Report

On the 10th of February I transmitted the following report of observations made by one of my assistants on Monday afternoon, 3 p.m., near Edwards Ferry:

Since my last observation I have discovered an increase of encampments in and about Leesburg, Va. They have commenced throwing up earth-works on the south side of Goose Creek and one mile and a half from the river.

No additional improvement has been made upon the old work that commands the ferry, and I think it is still unfinished.

The large fort west of Leesburg has been improved. It also appears that they have mounted some heavy guns. I could see no change about the works south of Leesburg. (I should judge that these were intrenchments.)

There are two large encampments (new) on the road running to the west from Leesburg, near the large stone house, which is, I think, one mile from town; also an encampment in the woods south of the large fort and west of the two encampments near the stone bridge.

On the north and south side of Leesburg I noticed an increase of encampments close to the town.

In and around the large fort west of Leesburg there is, I think, a regiment.

On Goose Creek, about three miles from the river, there are some encampments. I could not tell how many there were, as they are partly concealed by the woods.

About five miles to the southeast of Goose Creek and one mile from the river I observed large quantities of smoke rising from the woods.

To the rear of Ball's Bluff I observed a small camp (two or three companies).

Judging from the size and number of encampments, I should think there were from 10,000 to 12,000 troops opposite.

POOLESVILLE, *February* 21, 1862.

Professor T.S.C. Lowe's Official Report

T. S. C. LOWE:

I should like the balloon to be put in readiness to make an ascension as early as possible.

JNO. SEDGWICK,
Brigadier-General, Commanding Division.

On the 1st of March, by request of General Heintzelman, I was ordered to take a balloon to Pohick Church, on the Occoquan River, and the following are some of the reports made at that time:

POHICK CHURCH, *March* 5, 1862.

Captain MOSES,
Assistant Adjutant-General, Fort Lyon, Va.:

Have just made two ascensions with the balloon. It is fully inflated, and will take up two persons with all the ropes. If tomorrow is a fine day it would be a good time for the general to go up. I can see camp-fires on the Occoquan.

T. S. C. LOWE,
Chief Aeronaut, U. S. Army.

MARCH 6, 1862--11 *a.m.*

Brigadier-General MARCY,
Chief of Staff, Army of the Potomac:

I made two ascensions last evening. Saw fires at Fairfax Station; some on the road near the Occoquan. This morning cavalry scouts are visible on this side of the Occoquan below Sandy Run. There are five large smokes on the other side of the

49

Professor T.S.C. Lowe's Official Report

Occoquan, commencing at the ford below Wolf Creek and extending to the Potomac. Judging from appearances, compared with General Hooker's division, I should think their force inferior to his. The balloon at Budd's Ferry has been up all the morning at the same time with me. If the force here could be advanced across Pohick Creek on the heights, I should have no difficulty in getting very near the exact number of the enemy, as well as all of their fortified places.

We could also signal from one balloon to the other, which would be of importance to me.

I have sent for the balloon at Poolesville. Please inform me where to station it.

Your obedient servant,
T. S. C. LOWE,
Chief Aeronaut.

POHICK CHURCH, *March* 6, 1862.

Brig. Gen. R. B. MARCY,
Chief of Staff, Army of the Potomac:

GENERAL: I ascended at 5 this p.m. and remained up until 6 o'clock. It was calm and clear, and many of the enemy's camps were visible, and the smoke ascending straight gave a good idea of the enemy's position.

There are more smokes than usual at Fairfax Station, and a line of picket smokes extending southeast from there and nearly forming a junction with our lines running toward Springfield Station.

Heavy smokes (besides those seen in the morning) at Dumfries, Brentsville, Bradley's, and Manassas. General Heintzelman was here at 2 o'clock and went up twice.

I am greatly in need of that map that I spoke about yesterday

Professor T.S.C. Lowe's Official Report

to enable me to name places and distances more correctly. The one I have is small and inaccurate.

> Very respectfully, your obedient servant,
> **T. S. C. LOWE,**
> *Chief Aeronaut.*

POHICK CHURCH, *March 6, 1862.*

Capt. E. SEAVER,
In Charge of Balloon, Budd's Ferry, Md.:

I saw your balloon up this morning, but not this p.m. If tomorrow morning is calm, I shall ascend at 7 o'clock, or the first favorable spell. Do the same at your place, with one of your signal officers, that I may see if signals may be used at this distance.

> **T. S. C. LOWE,**
> *Chief Aeronaut.*

On the 7th General Berry, of General Heintzelman's command, ascended several times and discovered the evacuation of the Occoquan, which he reported to the latter officer. This was the first indication of the retirement of the enemy from Manassas.

I was personally absent in Washington, preparing a balloon to be taken to a point near Fairfax Court-House to watch for the evacuation, as it was somewhat expected, but for want of transportation I was unable to reach Fairfax until the 10th. To show with what esteem the commanding general held the operations of the aeronautic department, the following orders are submitted, which embrace all the items of interest up to the 1st of April, and it is believed that they indicate an appreciation of my services after an experience of the previous seven months:

WASHINGTON, *D. C., March 12, 1862.*

Professor T.S.C. Lowe's Official Report

Mr. T. S. C. LOWE,
Aeronaut, Army of the Potomac:

    SIR: You will make arrangements without delay to send to Fortress Monroe, Va., a balloon with all the requisite apparatus and materials for inflating it and making ascensions, and an aeronaut to manage the same.

By order of Major-General McClellan:
J. N. MACOMB,
Lieutenant-Colonel, Aide-de-Camp, in Charge of Balloons.

HALL'S HILL, *March 8, 1862.*

Professor LOWE,
National Hotel, Washington:

    I am authorized by General McClellan to call upon you for the balloon and inflating apparatus from Poolesville, and will be glad to have it here at as early an hour as possible, to take an observation a short distance in advance, where it will be well protected. Please reply what hour you will send it, as I desire to take an advantage of clear and calm weather. I wish the balloon but a few hours.

F. J. PORTER,
Brigadier-General.

HALL'S HILL, *March 9, 1862.*

Professor LOWE:

Professor T.S.C. Lowe's Official Report

Have your balloon out to Fairfax Court-House at as early an hour to-morrow as possible. Major Stone will give you all the facilities you desire. Show this to him.

By command of General F. J. Porter:
FRED. T. LOCKE,
Assistant Adjutant-General.

WASHINGTON, *D.C., March* 13, 1862.

Maj. Gen. J. E. WOOL,
Commanding Department of Virginia, Fortress Monroe, Va.:

GENERAL: By order of Major-General McClellan I send to you an aeronaut, Mr. E. Seaver, with a balloon and all necessary apparatus for making ascensions, who is instructed to report to you without delay.

I would very respectfully request that the aeronaut be furnished with such aid as may be required to manage the balloon to the best advantage, and trusting that by its means you will be able at all times to ascertain the position and movements of the enemy.

I remain, with respect, your obedient servant,
T. S. C. LOWE,
Chief Aeronaut, Army of the Potomac.

WASHINGTON, *D.C., March* 15, 1862.

Lieut. Col. J. N. MACOMB,
Aide-de-Camp and Major of Topographical Engineers:

Professor T.S.C. Lowe's Official Report

COLONEL: In accordance with orders, I proceeded yesterday from the Washington Navy-Yard to Budd's Ferry, and shipped the balloon and apparatus on board the steamer Hugh Jenkins, for Fortress Monroe, Va.

The dispatch which I had sent to Mr. Seaver to get the apparatus in his charge ready to move had not been received, and I found the balloon on the Virginia side of the river inflated, where it had been in use, and consequently my time was occupied during the entire night in getting the things together and shipping the same, in order that they might be on the way this morning, which I accomplished. On examination it was found impossible to turn the balloon barge until some repairs have been made to her rudder post, which got damaged during the late storm. I therefore sent the generator mounted upon wheels.

I sent Mr. Seaver to operate the balloon at Fort Monroe, with credentials, as Mr. Mason and Mr. C. Lowe did not arrive in time, they being occupied at Fairfax Court-House and Pohick Church, arranging apparatus for moving. I will send one of them to assist Mr. Seaver to-morrow or next day.

Very respectfully, your most obedient servant,
T. S. C. LOWE,
Chief Aeronaut.

SEMINARY, *March* 20, 1862.

Professor LOWE,
National Hotel, Washington:

I wish your balloon to embark with me at 9 to-morrow.

F. J. PORTER,
Brigadier-General, Headquarters near Seminary.

Professor T.S.C. Lowe's Official Report

HEADQUARTERS PORTER'S DIVISION,
March 21, 1862.

OFFICER IN CHARGE OF BALLOON PARTY:

SIR: You will prepare to embark this morning with this division. You will take three days' cooked provisions and three days' uncooked. You will be ready to march by 9 o'clock this morning.

By command of Brig. Gen. F. J. Porter:
FRED. T. LOCKE,
Assistant Adjutant-General.

HEADQUARTERS ARMY OF THE POTOMAC,
Near Alexandria Seminary, Va., March 22, 1862.

Mr. LOWE,
National Hotel, Washington, D.C.:

SIR: The commanding general directs that you make your arrangements to proceed to Fort Monroe with your balloons the same time that he moves, probably in the course of the following week.

The general will probably establish his headquarters on the steamer Commodore in a day or two.

I inclose your accounts approved, and with an order for its payment by Lieutenant-Colonel Macomb indorsed thereon.

Very respectfully, your obedient servant,
S. WILLIAMS,
Assistant Adjutant-General.

Professor T.S.C. Lowe's Official Report

HEADQUARTERS ARMY OF THE POTOMAC,
March 23, 1862.

T. S. C. LOWE,
Chief Aeronaut, National Hotel, Washington:

    The commanding general directs that you proceed with your balloons and apparatus to Fort Monroe, Va., and there await his further orders.

    It is understood that you have left a balloon with General Wadsworth, and that General F. J. Porter has one with him. Is this so? Please answer.

S. WILLIAMS,
Assistant Adjutant-General.

Professor T.S.C. Lowe's Official Report

Thaddeus Constantine Sobieski Lowe - Circa 1861

Field inflation of the balloon Intrepid before the Penninsula Campaign

Professor T.S.C. Lowe's Official Report

Members of T.S.C. Lowe's Balloon Corps utilizing the newly devised "twin" hydrogen generators to complete the inflation process

The inflation generators parked in the U.S. Army Supply Depot

Professor T.S.C. Lowe's Official Report

Ascension of the balloon "Intrepid" during the Battle of Fair Oaks

Lowe's balloon, "The City of New York," later renamed the "Great Western," to be used to a transatlantic flight

Professor T.S.C. Lowe's Official Report

T.S.C. Lowe

Leontine Augustine Gaschon Lowe

Professor T.S.C. Lowe's Official Report

Lowe's intended flight from Cincinnati

Two of the hydrogen gas generators assigned to each balloon for inflating on the battlefield.

Professor T.S.C. Lowe's Official Report

Note from President Abraham Lincoln to General Winfield Scott authorizing the formation of the Balloon Corps

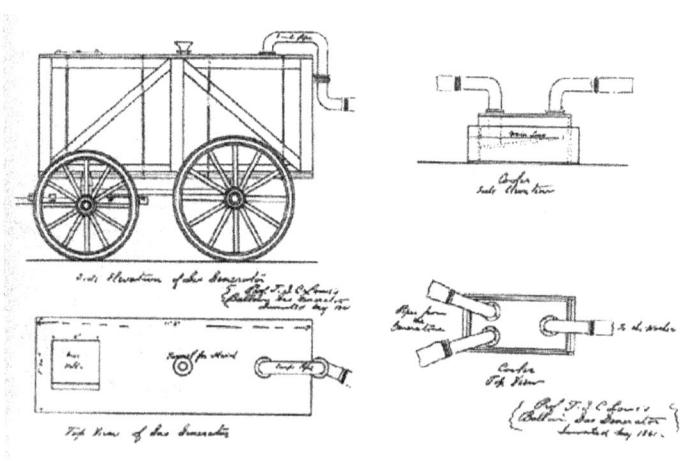

Schematics of the gas generators utilized by Lowe in the field

Professor T.S.C. Lowe's Official Report

Lowe's Balloon Corps personnel at the ready

T.S.C. Lowe in Camp - 1862

Professor Thaddeus Lowe's Official Report Part II

Professor T.S.C. Lowe's Official Report

Professor Thaddeus Lowe's Official Report Part II

(O.R.--SERIES III--VOLUME III [S# 124] CORRESPONDENCE, ORDERS, REPORTS, AND RETURNS OF THE UNION AUTHORITIES FROM JANUARY 1, 1861 TO DECEMBER 31, 1863.--# 12)

HEADQUARTERS ARMY OF THE POTOMAC,
March 23, 1862.

Prof. T. S. C. LOWE,
Chief Aeronaut, National Hotel:

The commanding general directs that on arriving at Fort Monroe you land all your balloons save one, which you will keep on board subject to his future directions.

S. WILLIAMS,
Assistant Adjutant-General.

On the 3d of April I received an order from General McClellan to accompany General Porter in His advance to Yorktown. On the following morning at 5 o'clock the division left Hampton and advanced as far as Cockletown, and on the 5th arrived in front of Yorktown. The aeronautic train, consisting of four army wagons and two gas generators, having to move in the rear, arrived a little after noon and were put in position for inflating the balloon. Our operations were impeded for an hour or

Professor T.S.C. Lowe's Official Report

more by our position being shelled by the enemy, but notwithstanding this the balloon was ready at 5.30 o'clock, and an observation was taken by an officer of the general's staff. At 3 o'clock the next morning I ascended and remained up until after daylight, observing the camp-fires and noting the movements of the enemy. On descending a messenger handed me the following order:

APRIL 6, 1862.

Professor LOWE:

The commanding general desires you to make an ascension as soon as you can. Look for the movement of wagons and teams; also where the largest number of men are.

Send word what is passing as soon as you can.

Very respectfully,
FRED. T. LOCKE,
Assistant Adjutant-General.

These observations being of great importance, I went to General Porter's tent and made my report, and requested that he should ascend that he might judge for himself of the number of the enemy and strength of their works. This he did, and remained up one hour and forty-five minutes at an elevation of 1,000 feet, and within a mile of the enemy's works. On descending, all the generals were called together and a council held. During the day several draughtsmen were sent up who sketched maps of the positions of the enemy, &c. In the afternoon the Count de Paris ascended with General Porter, and near sundown General Butterfield ascended to a height of 1,000 feet.

The observations and maps thus made were of the greatest importance, and readily enabled the commanding officer to decide what course he would pursue.

In the evening of the same day I received the following order from General McClellan:

Professor T.S.C. Lowe's Official Report

HEADQUARTERS ARMY OF THE POTOMAC,
April 6, 1862.

Professor LOWE:

General McClellan directs that you send a balloon to General Keyes' headquarters at Warwick Court-House as soon as possible.

By command of Major-General McClellan:
A. V. COLBURN,
Assistant Adjutant-General.

In compliance with this order I proceeded to Fortress Monroe to move another balloon to General Keyes' command, and left the one then inflated and in use before Yorktown in charge of the only assistant aeronaut I was then allowed, excepting one in charge of the balloon-boat at Fortress Monroe.

After stationing the balloon at Warwick Court-House (the train having to move over the worst roads I ever saw) I started on the night of the 10th for Yorktown. Our lines having been changed during my absence, I found myself, about 9 o'clock p.m., within the enemy's lines. I was not sensible of the danger I was in until I heard signals given by a low whistle, which I at once knew to be those of the rebels, and accordingly cautiously retraced my steps and spent the night at the camp of one of our advanced regiments. The next morning at daybreak I took the road to Yorktown, and at 6.30 I was surprised by the descent of a balloon very near me. On reaching the spot I found it to be the one I had left in charge of my assistant at Yorktown, and General Fitz John Porter the occupant. The gas had entirely escaped when the balloon reached the earth, from the fact that the general in his eagerness to come to the ground (on finding that the rope by which the balloon was let up had parted) had opened the valve until all the gas had escaped, and as the balloon was constantly falling the silk was kept extended, and presented so large a surface to the atmosphere that it served

Professor T.S.C. Lowe's Official Report

the purpose of a parachute, and consequently the descent was not rapid enough to be dangerous.

I would here remark that a balloon suddenly relieved of its gas will always form a half sphere, provided it has a sufficient distance to fall in to condense a column of air under it. A thousand feet would, I presume, be sufficiently high to effect this and to make the descent in safety.

On inquiring into the cause of the accident I found that Mr. Allen, the assistant in charge of the balloon, had used but one rope, as had been his idea of topical ascents, instead of three and sometimes four, as I always did, and that rope had been partially injured by acid which had accidentally got on it.

I found it difficult for a time to restore confidence among the officers as to the safety of this means of observation on account of this accident, but the explanations and the personal ascensions I made gradually secured a return of their favor, and on the 13th of April I received the following communication:

APRIL 13, 1862.

PROFESSOR: General Barnard is General McClellan's chief engineer, and is located in his camp. General McClellan is very anxious for him to have an ascension early in the morning, and General B. will be prepared to accompany your messenger, whom I beg of you to direct to wait to take General Barnard to the location of the balloon. I would ascend myself did not General B. wish and General McClellan wish him to go. General McClellan's camp is along the telegraph wire. Send the messenger to me if you do not know. I beg of you to give him a good and safe ascension.

Yours, truly,
F. J. PORTER.

P. S.--Send one of our men to rouse General B. at daylight, and wait to take him to your balloon. I think the best place is down the hollow where you were camped.

Professor T.S.C. Lowe's Official Report

On the following morning I called in person on General Barnard at daybreak and accompanied him to the balloon, when he ascended to an elevation of 1,000 feet and remained two hours. After breakfast he made two more ascents at different points, and expressed himself highly gratified with the information thus gained. From this time until the evacuation of Yorktown the balloons were kept in constant use, and reports were made by myself and many officers who ascended daily.

I regret that I have not more copies of reports, but as I had my camp at headquarters I usually made my reports verbally, assisted in my explanations by references to maps. Almost daily whenever the balloon ascended the enemy opened upon it with their heavy siege guns or rifled field pieces, until it had attained an altitude to be out of reach, and repeated this fire when the balloon descended, until it was concealed by the woods.

PORTER'S HEADQUARTERS,
April 29, 1862.

Captain MCKEEVER:

Please say to Professor Lowe, or his assistant, I would like to make an ascension as soon as the weather will permit, if they will notify me.

F. J. PORTER,
Brigadier-General.

Professor T.S.C. Lowe's Official Report

CAMP WINFIELD SCOTT,
Near Yorktown, Va., April 29, 1862.

Brig. Gen. S. VAN VLIET,
Chief Quartermaster, Army of the Potomac:

GENERAL: The commanding general directs that you cause to be issued to Professor Lowe, chief of balloon department of this army, such means of transportation and quartermaster's supplies as may be necessary to enable him to perform the duties with which he is charged.

Very respectfully, your obedient servant,
S. WILLIAMS,
Assistant Adjutant-General.

On the 3d of May I made a reconnaissance near Warwick Court-House and again before sundown before Yorktown, General McClellan and staff being on the spot; General Porter and myself ascended. No sooner had the balloon risen above the tops of the trees than the enemy opened all of their batteries commanding it, and the whole atmosphere was literally filled with bursting shell and shot, one, passing through the cordage that connects the car with the balloon, struck near to the place where General McClellan stood. Another 64-pounder struck between two soldiers lying in a tent, but without injury. Fearing that by keeping the balloon up the enemy's shots would do injury to the troops that were thickly camped there, General Porter ordered the balloon down. While making preparations to ascend again I received the following order:

YORKTOWN, *May* 3, 1862.

Professor LOWE:

Professor T.S.C. Lowe's Official Report

The general says the balloon must not ascend from the place it now is any more.

G. MONTEITH.

At about midnight, however, I was aroused by Captain Moses, of General Heintzelman's staff, who informed me that the general was apprehensive that the enemy were evacuating, from the fact of the constant cannonading, and that a heavy fire was also raging in Yorktown. I immediately ascended and saw that the fire was confined to one building or vessel near the wharf, and therefore I did not consider it a sufficient indication that they were evacuating, for if destruction of property was intended, they would burn their barracks, tents, wharves, store-houses, &c. I therefore considered the fire to be accidental.

I did not sleep any more, however, that night, and got the balloon ready for another ascension, which I made before daylight; but, as formerly, at this time in the morning I could see no camp-fires. As soon as it became a little lighter I discovered that the enemy had gone. This I immediately communicated to General Heintzelman, who on learning it ascended with me, satisfied himself of the fact, and reported it by telegraph to General McClellan, sending the message down from the balloon without descending. We then remained up and saw our troops advance toward the empty works, throwing out their skirmishers, and feeling their way as if expecting to meet an enemy. Of course we had no means of communicating to let our advance guard know where the enemy were, which we could see, as their rear guard was not more than one mile from Yorktown.

From the above facts it is fair to presume that the first reliable information given of the evacuation of Yorktown was that transmitted from the balloon to General McClellan by General Heintzelman and myself. Further proof of this, if necessary, will be found in General Heintzelman's report of the battle of Williamsburg, which I regret I have not at hand to quote from.

Professor T.S.C. Lowe's Official Report

I would also refer to the pamphlet written by Prince de Joinville, where in speaking of the evacuation of Yorktown and in other places he alludes to the ascensions of the balloon as an everyday occurrence in the Army of the Potomac for reconnaissances, and of their being frequently fired at by the enemy.

At about 7 o'clock the balloon was taken into Yorktown and observations made of the river for thirty miles. From the reports made that a number of vessels were in sight, our gun-boats were enabled to capture some and cause the destruction of many more.

To show how suddenly the enemy withdrew from Yorktown, I insert the following report to General Keyes, made verbally at the time and subsequently in writing:

ROPER'S MEETING-HOUSE, *May* 11, 1862.

Brig. Gen. E. D. KEYES,
Commanding Fourth Corps, Army of the Potomac:

GENERAL: In accordance with your request that I should give you a statement of the results of my observations from the balloon stationed at General Smith's division, near Warwick Court-House, on Saturday, May 3, I give the following: I ascended at noon, and remained at an elevation of nearly a thousand feet for one hour. Could see the rebel line of works and camps from York to James Rivers. At a point which I took to be Lee's Mill there seemed to be a large camp and earth-works as well as many others to the right and left. In several places there seemed to be gangs of men apparently throwing up earth-works. In addition to their barracks, many tents were visible, and, in fact, signs of evacuation were not visible. I reported the result of my observation to General McClellan on the same evening, and also to you at Brigadier-General Smith's headquarters at about 4 p.m. the 3d instant. On the following morning I ascended at a point near Yorktown and discovered that the enemy had left, and at 6 o'clock a portion of them were visible about two miles from Yorktown on the road to Williamsburg.

Professor T.S.C. Lowe's Official Report

Very respectfully, your obedient servant,
T. S. C. LOWE,
Chief Aeronaut.

It was known by all who had an opportunity of knowing that the enemy continued their works and kept up appearances until the night of the evacuation, and even kept their batteries firing until after midnight. Their barracks and tents, many of them new, were all left standing. Medical stores and ammunition (some destroyed and thrown into the river) were left, which it would seem would not have been the case if the evacuation had been long premeditated.

It is true army wagons were daily seen plying between Yorktown and Williamsburg, and so reported, but it was impossible to say which way they were loaded.

On the afternoon of the 4th I received orders to move everything pertaining to my department by water, with General Franklin's command. Judging from my orders, it would seem that the battle of Williamsburg was not expected.

The balloons were accordingly taken to West Point, and one was inflated on the balloon boat and used by General Franklin during his stay at that place, where reports were made to him of the position and movements of the enemy. After this we moved by water to White House Landing, the balloon boat being the first to land, and was even some distance ahead of the gun-boats, while the first night the balloon guard was the advance picket on the river bottom.

On the 18th of May I received orders to accompany General Stone-man, who was then some distance in the advance. We arrived near the Chickahominy on the morning of the 20th, and on the following morning, accompanied by General Stoneman, I ascended, and there had a distant view of Richmond, the general being the first to point out the city as we were rising. After ascertaining the location of the enemy, General Stoneman advanced his forces to Gaines' Hill, and there rested until the main portion of the enemy, which was still some distance in the rear, came up, while in the meantime the balloon was kept in constant

Professor T.S.C. Lowe's Official Report

use, and all the movements of the enemy were reported.

On the 25th of May the balloon proved of great advantage, and I copy the following memorandum from my notebook respecting the observations made:

GAINES' HILL, *May* 25, 1862.

This has been a fine and important day. General Stoneman ascended with me to an elevation of a thousand feet; had a splendid view of the enemy's country; discovered a force of the enemy near New Bridge, concealed to watch our movements. The general then took two batteries and placed them to the right and left of Doctor Gaines' house, and caused the enemy to retreat for at least a mile and a half, while he remained in the balloon with me, directing the commanders of the batteries where to fire, as they could not see the objects fired at. The general then went to Mechanicsville and drove the enemy from that position, while I remained up in the balloon to keep up appearances and to see if a larger force opposed him.

After descending, General Stoneman was heard to say, in the presence of several gentlemen, that he had seen enough to be worth millions of dollars to the Government.

It is certain that he is too keen an observer and too able an officer to be insensible of the advantages of so superior and accurate means of observation as that afforded by the balloon.

One of the principal objects of General Stoneman in driving the rebels from the banks of the Chickahominy was to enable him to move to Mechanicsville unnoticed, whereby he might surprise the enemy at that point, which he effectually accomplished by the aid of the balloon. He often availed himself of it by ascending personally, instead of trusting to some inferior officer who had no interest or reputation at stake. I had always noticed, moreover, that the general invariably pitched his tent where he could see the enemy himself.

On the occasion above alluded to the enemy were so concealed behind woods and hills that it was impossible to ascertain their positions in any other way than by ascending to a

Professor T.S.C. Lowe's Official Report

great elevation, and the artillery might have been fired a whole day without doing any injury, unless the proper range had been obtained.

A Richmond paper of May 26 contained the following item:

The enemy are fast making their appearance on the banks of the Chickahominy. Yesterday they had a balloon in the air the whole day, it being witnessed by many of our citizens from the streets and house tops. They evidently discovered something of importance to them, for at about 4 p.m. a brisk cannonading was heard at Mechanicsville and the Yankees now occupy that place.

On several other occasions the Richmond papers correctly described the various ornaments painted on the balloons, as seen with telescopes from the city.

On the 26th and 27th I received the following orders:

HEADQUARTERS ARMY OF THE POTOMAC,
INSPECTOR-GENERAL'S DEPARTMENT,
May 26, 1862.

Professor LOWE:

SIR: I am instructed by Brigadier-General Marcy, chief of staff, to direct you to move your balloon, &c., with as little delay as possible, to Brigadier-General Stoneman's headquarters, at Mechanicsville.

You are directed after each ascent to make a written report to the headquarters of the result of your observations.

I am, sir, very respectfully, your obedient servant,
D. B. SACKET,
Inspector-General, U.S. Army.

Professor T.S.C. Lowe's Official Report

HDQRS. TOPOGRAPHICAL ENGINEERS, *ARMY OF THE POTOMAC,*
May 26, 1862.

Prof. T. S. C. LOWE,
Chief Aeronaut, Army of the Potomac:

SIR: The balloon department has been placed under my direction by Special Orders, No. 157, May 25. Understanding that there are several balloons in your charge, you will immediately establish them in the following positions, viz, near Mechanicsville, General Stoneman commanding; near the Seven Pines, on the road from Bottom's Bridge to Richmond, about six miles from the bridge, General Keyes commanding, and in the vicinity of New Bridge, near the general headquarters.

Very respectfully, your obedient servant,
A. A. HUMPHREYS,
Brig. Gen. and Chief Topographical Engineers.

HDQRS. TOPOGRAPHICAL ENGINEERS, *ARMY OF THE POTOMAC,*
May 27, 1862.

Prof. T. S. C. LOWE,
Chief Aeronaut, Army of the Potomac:

DEAR SIR: The general commanding desires-first, that balloon ascensions be made as frequently as is practicable at each balloon station and that full reports of one results of the observations be transmitted at once to these headquarters; second, that no persons be permitted to ascend in the balloon with the exception of the general in command at the position which the balloon occupies, and those authorized by him; third, that

Professor T.S.C. Lowe's Official Report

newspaper correspondents and reporters be in no case permitted to ascend.

Very respectfully, your obedient servant,
A. A. HUMPHREYS,
Brig. Gen., Chief of Topographical Engineers, Army of the Potomac.

It will be seen from the following dispatches that the enemy improved every opportunity to fire at the balloon. On this occasion I ascended to a high altitude, and before I descended I had the balloon moved considerably to one side, so that the subsequent firing was out of range, and thus, by changing my location, prevented the enemy from having a good mark to fire at.

MAY 27, 1862.

Gen. A. A. HUMPHREYS:

Ascended at 4.45 p.m. one mile from Mechanicsville and, I should judge, four miles from Richmond, in an air line. At 5 o'clock three batteries opened upon me, firing many shots, some falling short and some passing beyond the balloon and one over it, while it was at an elevation of 300 to 400 feet. A battle is going on about four miles distant; heavy cannonading and musketry. I will go up again and report.

T. S. C. LOWE.

MAY 27, 1862.

Brigadier-General HUMPHREYS,
Chief of Topographical Engineers:

Professor T.S.C. Lowe's Official Report

GENERAL: I made my second ascent at 5.30 p.m., and remained up until 6.45 p.m. Richmond and vicinity are much more distinct from this point, and I was able to discover with ease the exact position of the enemy. The heaviest camps seem to be near the banks this side of James River and a little to the left of Richmond. The next heaviest are to the right of Richmond on the road from Mechanicsville. There are also several smaller on the first heights opposite Mechanicsville, and several batteries stationed there, some of which I saw put in position while in the balloon, besides those that fired at me.

The heights opposite New Bridge for two miles each way seem to be entirely unoccupied, except by the enemy's pickets.

No earth-works of any description are visible, although the country is tolerably clear from woods on the Mechanicsville road, and if there are earth-works on this side they are very near the city and behind the last line of woods.

In the northwest from where the balloon is, and about ten miles distant, there was heavy smoke.

To the north, near the Pamunkey River, was the heavy cannonading and musketry, but the distance and heavy woods prevented me from seeing the detail movements. The enemy in and around Richmond are apparently very strong in numbers.

Very respectfully, your obedient servant,
T. S. C. LOWE,
Chief Aeronaut.

BALLOON CAMP,
Near Mechanicsville, May 29, 1862--9.30 a.m.

Brig. Gen. A. A. HUMPHREYS,
Chief of Topographical Engineers, Army of the Potomac:

GENERAL: I ascended at 7.30 o'clock this a.m., near New Bridge; could discover no change in the position of the enemy in that vicinity. I then came to this point to get another view, which I have just obtained, and find the enemy quite opposite

Professor T.S.C. Lowe's Official Report

Mechanicsville.

    A battery consisting of several guns is in position near the road on the opposite heights. There are troops lying in the shade of the woods along the whole line from below New Bridge to some distance above this point, the greatest number, however, opposite this point.

    I have now on hand material sufficient to keep the two balloons in operation for about one week only.

Very respectfully, your obedient servant,
T. S. C. LOWE,
Chief Aeronaut, Army of the Potomac.

    From 11 o'clock until dark on the 29th of May the enemy commenced to concentrate their forces in front of Fair Oaks, moving on roads entirely out of sight of our pickets, and concealing themselves as much as possible in and behind woods, where none of their movements could be seen, except from the balloon. The following is one of my reports on that day:

BALLOON CAMP,
Near New Bridge, May 29, 1862.

Brig. Gen. A. A. HUMPHREYS,
Chief of Topographical Engineers:

    GENERAL: My last ascent was made at sundown, which discloses the fact that the enemy have this afternoon established another camp in front of this point in the edge of the woods to the left of the New Bridge road and on a line with the permanent camp about one mile and a half to two miles from the opposite heights. They seem to be strengthening on our left, opposite this place.

Very respectfully, your obedient servant,
T. S. C. LOWE.

Professor T.S.C. Lowe's Official Report

P. S.--My last dispatch dated 1.30 o'clock ought to have been 3. My watch had stopped.

LOWE.

On that night or the following morning General McClellan ordered the reserves to be moved up to support General Heintzelman in case of an attack, which took place just as this was accomplished. Had not our forces been concentrated it is very evident that our left, or that portion of our army beyond the Chickahominy, would have been driven back, and in consequence the whole army routed.

I think that I have reason to presume that the cause of this favorable movement of our troops was mainly due to my report that the enemy were moving down and strengthening in front of Fair Oaks.

On the 31st of May, at noon, I ascended at Mechanicsville, and discovered bodies of the enemy and trains of wagons moving from Richmond toward Fair Oaks. I remained in the air watching their movements until nearly 2 o'clock, when I saw the enemy form in line of battle, and cannonading immediately commenced. Not having any telegraphic communication here, I dispatched one of my assistants with a verbal message, and, to make the matter doubly sure, I sent the following written dispatch after reaching Doctor Gaines' house forty-five minutes later, and still another at 4.30 p.m.:

DOCTOR GAINES' HOUSE, *May* 31, 1862.

General MCCLELLAN:

I descended at 2 o'clock from near Mechanicsville. The position of the engagement is about four or five miles from New Bridge in a southerly direction. Could see troops moving toward the firing from our left of Richmond, and a long wagon train also moving in that direction.

The enemy on our right seem to remain quiet. Quite a large

Professor T.S.C. Lowe's Official Report

reserve are in the edge of the woods about one mile and a half from the heights on the road from New Bridge. I will ascend from this point as soon as the wind lulls.

Your very obedient servant,
T. S. C. LOWE.

MAY 31, 1862--4.30 *p.m.*

Brigadier-General MARCY,
Chief of Staff:

There are large bodies of troops in the open field beyond the opposite heights on the New Bridge road. White-covered wagons are rapidly moving toward the point of the engagement with artillery in the advance. The firing on our left has ceased.

T. S. C. LOWE,
Chief Aeronaut.

On receipt of the above information General McClellan sent express orders to General Sumner to have the bridge across the Chickahominy completed as soon as possible, and to cross with his corps at the earliest possible moment and support General Heintzelman. This was accomplished just in time, for it is asserted upon good authority that if General Sumner had been one or two hours later the day would have been lost.

Is it not probable, to say the least, that my reports from the balloons caused the completion of this bridge two hours sooner than it would otherwise have been done? In reference to this point I would refer to the Prince de Joinville's narrative of the Peninsular Campaign, where in speaking of the battle of Fair Oaks he says that "there was some doubt whether the enemy were making a real attack, or whether it was merely a feint; *but this doubt was soon*

Professor T.S.C. Lowe's Official Report

removed by reports from the aeronauts, who could see heavy columns of the enemy moving in that direction."

On the following morning I ascended at 4 a.m., but owing to fog I was unable to see anything until after 6 o'clock, and at 7 o'clock I sent the following dispatch by telegraph from the balloon.

Many dispatches were sent in this way, copies of which were not preserved:

> NEAR DOCTOR GAINES' HOUSE,
> *June* 1, 1862--7 *a.m.*

Brigadier-General HUMPHREYS, *or*
General MARCY,
Chief of Staff:

I have just obtained a splendid observation from the balloon. I find the enemy in large force on the New Bridge road, about three miles this side of Richmond. In fact, all of the roads that are visible are filled with infantry and cavalry moving toward Fair Oaks Station. There is also a large force opposite here, and in the same position that they were yesterday, but not in motion. I can see smoke in the woods where the firing ceased last night. I hear no firing at the present. In the immediate vicinity of the heights opposite here there are nothing but pickets visible.

> T. S. C. LOWE,
> *Chief Aeronaut.*

I am satisfied from what I heard on the previous evening that an attack by the enemy on the next morning was not expected. The above dispatch, therefore, giving timely notice that the enemy did really intend making a more severe attack than even that of the previous day, must certainly have been of the greatest importance, and gave our forces an opportunity of preparing for a vigorous defense.

I would here remark that of all the battles I have witnessed,

Professor T.S.C. Lowe's Official Report

that of Fair Oaks was the most closely contested and most severe, and the victory, in my opinion, was due to the valor and skill of General Heintzelman, who nobly sustained himself against great odds in favor of the enemy.

To the following reports I would call especial attention, as they speak for themselves.

The following order from General Humphreys was received one hour after my first report:

JUNE 1, 1862--6.45 *a.m.*

Professor LOWE:

Have you been able to ascend this morning? Your balloon should be in connection by telegraph, and messages should be sent constantly--at least every fifteen minutes. The balloon must be up all day. The balloon at Mechanicsville should likewise be sent up at once, and remain up all day.

Same reports must be made from it as from the balloon at Doctor Gaines'.

A. A. HUMPHREYS,
Brigadier-General.

BALLOON CAMP,
Doctor Gaines' House, June 1, 1862.

Brigadier-General HUMPHREYS, *or*
General MARCY,
Chief of Staff:

The reserve of the enemy are considerably strengthened on the New Bridge road, and troops are still moving that way from Richmond; they do not seem to be gathering in any great numbers

Professor T.S.C. Lowe's Official Report

on the immediate heights along the Chickahominy. Our supports, with army wagons, are in a southeast direction from here, advancing, and about three miles from the fire of yesterday. Musketry is in constant operation in the same direction as yesterday. The banks of the Chickahominy are overflowed as far as can be seen.

Respectfully, your obedient servant,
T. S. C. LOWE,
Chief Aeronaut.

BALLOON CAMP,
Near Doctor Gaines' House, June 1, 1862--11 a.m.

Brigadier-General HUMPHREYS, *or*
General R. B. MARCY,
Chief of Staff:

 My ascent and observations just completed show the firing of the enemy to be in the same position. The road in the rear of the firing is filled with wagons and troops. About two miles still farther to the rear of Fair Oaks Station, and on the Williamsburg stage road, Charles City road, and Central road, are also large bodies of troops; in fact, I am astonished at their numbers compared with ours, although they are more concentrated than we are. Their whole force seem to be paying attention to their right. A regiment has just marched to the front, where we are preparing a crossing. Their large barracks to the left of Richmond is entirely free from smoke, and, in fact, the whole city and surroundings are nearly free from smoke, which enables me to see with distinctness the enemy's earth-works. Quite a large body of troops are on the other side of the river, about two miles from here, to our left.
 The weather is now calm, and an excellent opportunity is offered for an engineer officer to accompany me.
 The balloon at Mechanicsville is constantly up.

Professor T.S.C. Lowe's Official Report

Your very obedient servant,
T. S. C. LOWE.

JUNE 1, 1862.

Professor LOWE:

    Direct your attention to a force said to be approaching toward our left, apparently to attack the working parties at the bridge below New Bridge. It is said a gun is planted to strike the bridge. Send me intelligence by bearer and at once communicate to me or General ----- , when present, what is passing.

J. H. MARTINDALE,
Brigadier-General, in Charge of Porter's Division.

JUNE 1, 1862--12.15 *p.m.*

General MARTINDALE:

    About one hour ago a full regiment moved up into the woods toward where our left crossing is being made. I have seen no artillery moved up, nor can I see any from here. I think, however, there is artillery in the woods.

Very respectfully,
T. S. C. LOWE,
Chief Aeronaut.

HEADQUARTERS OF GENERAL MCCLELLAN,
June 1, 1862.

Professor T.S.C. Lowe's Official Report

Professor LOWE:

The enemy has been repulsed wherever he attacked. Watch the motions of the enemy and his wagons and see where goes the force before Mechanicsville.

R. B. MARCY.

Professor LOWE:

Can you see General Sumner's corps near the line of railroad about four miles from the Chickahominy? Was the train of our wagons you saw going toward Richmond or toward James River? Can you see the gun-boats on James River? Which direction does the smoke run?

R. B. MARCY,
Chief of Staff.

JUNE 1, 1862.

At 11 o'clock could see what I understood was General Sumner's corps near the line of railroad, but not more than two miles from the Chickahominy. The wagons I saw were moving toward James River. They had not reached the road to Richmond.

I cannot see the gun-boats, but can see heavy smoke arising from the valley at two points, and hear heavy reports from cannon. The enemy's reserves seem to be stationed at present in all the roads.

T. S. C. LOWE,
Chief Aeronaut.

Professor T.S.C. Lowe's Official Report

The following were answers to questions asked by General Porter:

JUNE 1, 1862--3 *p.m.*

Brig. Gen. F. J. PORTER:

The enemy remains quiet opposite New Bridge. There are infantry and a battery of artillery near the river, where our left column is preparing to cross. The wind is now too high to get a view opposite Mechanicsville, and I am not in immediate communication with the balloon there. By the appearance of the smoke when up I would say that we hold our ground, and more too. The Chickahominy is fast rising; in front of this point the whole fields resemble a lake.

The enemy's wagons also seem to be stationary opposite here.

Very respectfully,
T. S. C. LOWE,
Chief Aeronaut.

BALLOON IN AIR, *June* 1, 1862---6.30 *p.m.*

R. B. MARCY,
Chief of Staff:

Last firing is two miles nearer Richmond than this morning. Camp-fires around Richmond as usual, showing that the enemy are back. General Humphreys and staff are now up, and will endeavor to ascertain fully and answer all your questions.

T. S. C. LOWE,
Chief Aeronaut.

Professor T.S.C. Lowe's Official Report

JUNE 1, 1862--7 *p.m.*

Brigadier-General MARCY,
Chief of Staff:

General Humphreys and self have just descended. The enemy is still in the field opposite here, and their works are visible all along the Williamsburg and New Bridge roads to Richmond. Their barracks, which were this morning deserted, are now occupied. I can see no wagons moving in any direction. Brigadier-General Humphreys will give you a full account of the last observation. I will ascend again at daybreak to-morrow.

Your very obedient servant,
T. S. C. LOWE,
Chief Aeronaut.

BALLOON CAMP, *NEAR MECHANICSVILLE,*
Sunday Morning, June 1, *1862---8.20 a.m.*

Large force in front of New Bridge. Do not think there is a very large force in front of Mechanicsville. The rebels have struck their tents in front of the above-named place (Mechanicsville).
10.45 a.m.--The rebels are moving a brigade out of Richmond in the direction of New Bridge.
11.10 a.m.--The brigade that I saw moving out of Richmond at 10.45 a.m. seems to be a very large one. They are followed by a train, consisting of twenty-four wagons, and have just entered the woods, which carries them out of my sight. Think they are going in the direction of New Bridge.
The troops that were in front of New Bridge have fallen back under cover of the woods.
(The above are copies of Major Webb's dispatches to General Marcy, as far as I can remember.) Major Webb was up in the balloon from 8 a.m. till 11.20 a.m.

Professor T.S.C. Lowe's Official Report

JAMES ALLEN,
Assistant Aeronaut.

JUNE 2, 1862--5.25 *a.m.*

Brigadier-General HUMPHREYS, *or*
General MARCY,
Chief of Staff:,

    I ascended at 4.45 this a.m. Found the enemy in full force opposite this point, with their horses harnessed to their artillery. I observed their movements for half an hour; saw mounted pickets to the extreme left of the large field opposite the point where we are preparing a crossing. To the right, opposite Mechanicsville, the enemy have two large camps, and all along their line there are appearances of lively movements.
    In fifteen minutes from the time of my ascent a battery of six guns left the farther side of the field, on the New Bridge road, and came to the heights opposite here and covered themselves in the woods, just one mile and three-quarters from this point. I am confident from the present movements that they intend to intercept our crossing the river. The weather at present is calm, and a good opportunity for some officer to ascend in the Mechanicsville balloon before the storm, which I think is near at hand. I would suggest Major Webb, as he is accustomed to the balloon.

Your very obedient servant,
T. S. C. LOWE,
Chief Aeronaut.

JUNE 2, 1862--10.15 *a.m.*

Professor T.S.C. Lowe's Official Report

Brigadier-General HUMPHREYS, *or*
General MARCY,
Chief of Staff:

    The enemy remain quiet and in the same position as reported at 8.15. Large numbers are at work throwing up earth, as before, opposite General Smith's headquarters.
    Lieutenant-Colonel Palmer could not stand an ascension, owing to vertigo.

T. S. C. LOWE,
Chief Aeronaut.

McCLELLAN'S HEADQUARTERS, *June* 3, 1862.

Professor LOWE:

    It is reported that the enemy in force is advancing on our troops to the left, in front of Sumner and Heintzelman. Please make an ascension as soon as practicable and inform me what you discover in that direction, and make frequent ascensions afterward.

R. B. MARCY,
Chief of Staff.

DOCTOR GAINES' HOUSE, *June* 3--2.45 *p.m.*

Brigadier-General MARCY,
Chief of Staff:

    Just as I received your dispatch General Barnard arrived and remained up about twenty minutes. I have just descended myself. I

Professor T.S.C. Lowe's Official Report

could see no additional troops at the point you inquire about. There have been troops for the past three or four days on the New Bridge road about one mile beyond Doctor Garnett's house, or red brick house opposite here, and daily moving about in regiments forward and back as a picket force. I can discover no new movements of the enemy to-day.

T. S. C. LOWE,
Chief Aeronaut, Army of the Potomac.

    General Barnard made very frequent ascensions during the whole time our army lay before Richmond, and from observations thus taken he was better enabled to locate earth-works, &c., of which many were constructed.
    The following are dispatches without dates, which I take the liberty of adding, as they were accidentally omitted from the copies I retained.
    Before the battle of Fair Oaks:

Brig. Gen. A. A. HUMPHREYS,
Chief of Topographical Engineers:

    GENERAL: I ascended at sunrise this morning. The enemy's line of pickets in front of this point (Doctor Gaines' house) remains, as usual, from one-half to three-quarters of a mile from the Chickahominy, about one mile and a half from the heights opposite this point, and on the road from New Bridge still remains the camp noticed in my first ascent, some days since, apparently without any increase. Directly south of this point, about five miles, is a tolerable-size-d camp smoke, and I should judge about three miles and a half in advance of the main camp of General Keyes.
    The city of Richmond was entirely enveloped in smoke. The balloon at Mechanicsville was also up at the same time with me. I will make an ascent from Mechanicsville as soon as the atmosphere clears.

Professor T.S.C. Lowe's Official Report

Very respectfully, your obedient servant,
T. S. C. LOWE,
Chief Aeronaut.

The three following reports were made after the battle of Fair Oaks:

Brig. Gen. A. A. HUMPHREYS:

GENERAL: I have just completed another observation from the balloon. About three-quarters of a mile from the heights opposite here, and about two miles and a half from this point, are about six regiments of infantry. Trees have been felled beyond them, so that I can now see another small field beyond where trees were standing this morning. There is heavy smoke now rising, as though underbrush were burning. I will watch their operations and report.

T. S. C. LOWE.

7.15 A.M.

Brigadier-General HUMPHREYS, *or*
General R. B. MARCY,
Chief of Staff:

The enemy remain the same opposite this point. I can see through a small open space in the woods, on what I think is the Williamsburg road, troops moving toward the late scene of action, but not in great numbers, however.

Very respectfully, your obedient servant,
T. S. C. LOWE.

Professor T.S.C. Lowe's Official Report

8.15 A. M.

Brigadier-General HUMPHREYS, *or*
General R. B. MARCY,
Chief of Staff:

The atmosphere is now quite clear. The troops still remain quiet opposite here. On the heights opposite General Smith's headquarters and on the left-hand side of the New Bridge road, going to Richmond, the enemy are throwing up earth. Many army wagons are remaining stationary in that direction and horses grazing.

Respectfully,
T. S. C. LOWE,
Chief Aeronaut.

CAMP NEAR DOCTOR GAINES' HOUSE,
June 3, 1862--5 *a.m.*

Brigadier-General HUMPHREYS:

I ascended this morning at an altitude of 900 feet just before 5 o'clock, but found the atmosphere so thick with mingled smoke and fog that only a few places were visible. The enemy opposite this point remain the same as yesterday, and along the heights for two miles up nothing is moving on the roads.

T. S. C. LOWE,
Chief Aeronaut, Army of the Potomac.

HEADQUARTERS ARMY OF THE POTOMAC,
June 7, 1862.

Professor T.S.C. Lowe's Official Report

Professor LOWE :

You will please allow Mr. Babcock to make ascensions in your balloon whenever it is convenient. He is making maps and desires to make observations.

R. B. MARCY,
Chief of Staff.

BALLOON CAMP,
Doctor Gaines' House, June 7, 1862.

Brigadier-General HUMPHREYS, *or*
General *MARCY,*
Chief of Staff:

I ascended at 6 o'clock and remained up in all about one hour. The enemy appears to be in larger force on our left than at any other point. Our advance and the enemy's artillery are less than one mile from each other. The artillery that I refer to is about half a mile to the left of the New Bridge road, in the field and behind the woods on the opposite heights, with horses attached; there is more in the rear, with horses picketed. Their picket-line is not so far advanced as formerly. Several squads of cavalry were visible along the opposite heights. There are large camp smokes opposite Mechanicsville and beyond, but the dense haze prevents me at this time from observing details.

The Intrepid will lift three persons and ropes, and there will be an excellent opportunity for engineers to ascend. I will go up early in the morning again.

Your very obedient servant,
T. S. C. LOWE,
Chief Aeronaut.

Professor T.S.C. Lowe's Official Report

BALLOON CAMP,
Doctor Gaines' House, June 9, 1862.

Brigadier-General HUMPHREYS, *or*
General *MARCY,*
Chief of Staff:

I ascended at sundown this p.m. and find the enemy's camps located about the same as they have been for the past four or five days.

Two sections of a battery, of three guns each, are stationed in the field (with horses attached) about three-quarters of a mile southeast from Doctor Garnett's house. Two other batteries are stationed near Old Tavern. Very heavy camps are still beyond and to the right toward Richmond. There are also three distinct camps extending from Widow Price's to Doctor Friend's, on a road this side of the New Bridge road.

Pickets are visible near General Smith's advance, but no fires are built. The enemy's smokes immediately in front of the late battle-grounds are very light. Owing to the lateness of the hour before I could ascend, in consequence of the heavy winds, I was unable to finish my observation to the right, but will ascend as often as possible.

Your very obedient servant,
T. S. C. LOWE,
Chief Aeronaut, Army of the Potomac.

BALLOON CAMP,
Doctor Gaines' House, June 10, 1862--4.30 *p.m.*

Brigadier-General HUMPHREYS,
Chief of Topographical Engineers:

Professor T.S.C. Lowe's Official Report

GENERAL: I ascended at 3.45 this p.m., but have nothing new to report. The enemy remain about as usual. It would be a good time for some one to ascend at Mechanicsville, but I am not able to ride there myself, and Mr. Allen is quite ill.

The atmosphere is quite clear, but the earth is heavily shaded by clouds.

Your very obedient servant,
T. S. C. LOWE,
Chief Aeronaut.

HEADQUARTERS FIFTH ARMY CORPS,
June 11, 1862.

Professor LOWE:

SIR: The commanding general desires you to make an ascension this evening, if but for a few moments, to try if you can see anything of a large body of the enemy, said to be in the vicinity of Old Tavern, near Mrs. Price's house.

Very respectfully, your obedient servant,
F. T. LOCKE,
Assistant Adjutant-General.

BALLOON CAMP,
Near Doctor Gaines' House, June 12, 1862.

Brigadier-General HUMPHREYS, *or*
General MARCY,
Chief of Staff:

Professor T.S.C. Lowe's Official Report

I ascended at about sundown this p.m. The atmosphere very hazy beyond a distance of three miles. Could see no movements of the enemy. Their camps and camp-fires remain the same as usual.

Your obedient servant,
T. S. C. LOWE,
Chief Aeronaut.

GENERAL MCCLELLAN'S HEADQUARTERS,
June 13, 1862.

General F. J. PORTER:

Order Lowe to make frequent ascensions and report everything.

R. B. MARCY,
Chief of Staff.

The general wants you to look both ways--up and down the river and toward Mechanicsville. I send you two orderlies. Keep them till dark.

Yours,
F. T. LOCKE,
Assistant Adjutant-General.

BALLOON CAMP,
Near Doctor Gaines' House, June 13, 1862--6.15 *a.m.*

Professor T.S.C. Lowe's Official Report

Brigadier-General HUMPHREYS, *or*
General MARCY,
Chief of Staff:

    I ascended at 5.15 this a.m. and remained up one hour. The cannonading during the time I was up was from James Garnett's house (according to Allen's map), and directed to one of our camps to the left of General Smith's. Owing to the dense fog and smoke a view of all the roads could not be obtained, but on those that were visible I could see no movements whatever. I will ascend again as soon as the fog clears a little.

Respectfully,
T. S. C. LOWE.

NEAR DOCTOR GAINES' HOUSE,
June 13, 1862--8 *a.m.*

Brig. Gen. A. A. HUMPHREYS, *or*
General MARCY,
Chief of Staff:

    I have just completed another observation from the balloon. The enemy's artillery remains at the same point (James Garnett's), and, with the exception of two or three squadrons of cavalry and the usual picket, there are no other troops in position or on the visible roads. During the time of my observation the most of the enemy's shots fell short. There was no response from our side during the time.

Respectfully,
T. S. C. LOWE,
Chief Aeronaut, Army of the Potomac.

Professor T.S.C. Lowe's Official Report

HEADQUARTERS FIFTH ARMY CORPS,
June 13, 1862.

Professor LOWE,
Balloon Corps:

Large bodies of the enemy are reported to be moving with baggage wagons and ambulances toward our left. The commanding general desires you will make ascensions as often as practicable, observe their movements, and send up the information to him A dispatch sent to General Morell will be forwarded by him to these headquarters.

Very respectfully, your obedient servant,
FRED. T. LOCKE,
Assistant Adjutant-General.

Every few days after the battle of Fair Oaks alarming reports were circulated that the enemy in large force was moving to different points to make an attack, as will be seen by the above and previous orders, although many more were sent verbally. The balloon was always called into requisition to ascertain the truth of these reports, and in almost every instance our troops, who would otherwise have been compelled to lie upon their arms for hours and perhaps days, in addition to other exposure consequent upon building earth-works, roads, bridges, &c., were allowed to return to their quarters on receiving a report from the balloon that the enemy was quiet. It often seemed to me that these false reports were circulated expressly to annoy and weary our forces, and so reliable did they sometimes appear that on several occasions I was required to take up a staff officer and point out to him the location of the enemy before our generals could be satisfied.

JUNE 13, 1862--8.15 *p.m.*

Professor T.S.C. Lowe's Official Report

Brigadier-General MARCY,
Chief of Star:

My assistant at Mechanicsville reports that he has taken several observations this afternoon, and from appearances of smoke and troops he is of the opinion that the force opposite Mechanicsville is considerably strengthened.
I ascended from this point since my last dispatch and remained up until dark, but have nothing new to report.

Respectfully,
T. S. C. LOWE.

The following reports of June 14 were of the greatest importance, and gave the commanding general timely notice of the intentions of the enemy and enabled him to use his facilities to the best advantage. Knowing that the enemy could, after a few days' work, fortify themselves sufficiently to hold our forces in check with a portion of their army, until the remainder would be at liberty to operate in another direction, General McClellan could make his final attack then before the enemy were any stronger, or he could fortify himself, or prepare for a retreat, or change of base, just as his facilities would permit. At all events, about two weeks later it proved that the enemy was so fortified that they held their position with but a small portion of their force, while the main body of their army was thrown against our right, which they overpowered and compelled the retreat to James River.

BALLOON CAMP,
Near Gaines' House, June 14, 1862--9.30 *a.m.*

Brigadier General MARCY,
Chief of Staff:

GENERAL: I ascended at 8 and remained nearly one hour at an elevation of 1,000 feet. It was perfectly calm and many fields and camps were visible that I have not been able to see for a

Professor T.S.C. Lowe's Official Report

number of days past. In almost every field and on all available hills the enemy have large working parties throwing up earth-works and digging rifle-pits.

The camps and tents about Richmond seem to be much increased since my last good view beyond the woods. I can now count ten distinct earth-works around Richmond and can see embrasures in most of them, but cannot distinguish whether they have guns mounted in them or not. I am now marking upon the map the positions as near as possible of the earth-works now building, and will send it in to-day.

Your very obedient servant,
T. S. C. LOWE,
Chief Aeronant.

BALLOON CAMP, *June* 14, 1862.

Brigadier-General MARCY,
Chief of Staff, Army of the Potomac:

GENERAL: Accompanying this note is a map with some of the most important earth-works represented, and in the right place, as near as I can get them according to the map. There are other places where earth has been thrown up, but I shall have to ascend again to a high altitude in order to locate them. The work that commences at Widow Price's house runs to the woods a little to the right of Old Tavern, and on the farther end I should judge that 500 or more persons were at work this morning. There is also a small work to the right of the house, running into the woods.

The numbers upon the map are for the purpose of explaining the various points better when telegraphing from the balloon. Please preserve it for that purpose.

I am greatly in need of a good field glass for the Mechanicsville balloon. If one can be obtained will you please send it by the orderly, and greatly oblige,

Professor T.S.C. Lowe's Official Report

Your very obedient servant,
T. S. C. LOWE,
Chief Aeronaut, Army of the Potomac.

BALLOON CAMP,
Near Doctor Gaines' House, June 14, 1862--6.15 *a.m.*

Brigadier-General MARCY,
Chief of Staff:

    GENERAL: I remained in the air from 5 to 6 o'clock this morning. There appears to be no movements of the enemy upon any of the roads at this time. Many camp-fires were built during the time I was up, showing the enemy in the same position as yesterday. The artillery that was at James Garnett's house yesterday is not in sight this morning.

Your obedient servant,
T. S. C. LOWE.

BALLOON CAMP,
Near Doctor Gaines' House, June 16, 1862.

Brig. Gen. R. B. MARCY,
Chief of Staff, Army of the Potomac:

    GENERAL: The first ascension that I was able to make to-day was at 3.30 p.m.
    The enemy are still hard at work on their intrenchments all along their line.
    The work in front of Widow Price's extends farther along to the right than I at first supposed, as I can see by breaks in the

Professor T.S.C. Lowe's Official Report

woods when at a high altitude. It also runs some distance to the left and masked by bushes.

After remaining up nearly one hour Colonel Alexander ascended. I then went to Mechanicsville and had a fine view from that point. The enemy there appeared to be more in force immediately opposite Meadow Bridge than between Mechanicsville and Richmond.

There are two works in sight from the upper balloon---one near Caxton's, or No. 16, and another at 21, as marked on the map that I sent you. Much the largest force, however, and the most work going on, is in front of our left.

While up at Mechanicsville I saw what appeared to be two regiments moving on the New Bridge road--from figure 7 toward Thorn's, with thirteen covered wagons in the rear. I then came to this point and saw them come in near Old Tavern. There are several pieces of artillery visible near James Garnett's house. I will have a balloon in operation as soon as possible near headquarters.

Your very obedient servant,
T. S. C. LOWE,
Chief Aeronaut, Army of the Potomac.

BALLOON CAMP,
Near Doctor Gaines' House, June 17, 1862.

Brig. Gen. R. B. MARCY:

GENERAL: I took an observation this morning at 7 o'clock. Found the enemy still busy at work on their trenches. The work in front of Mrs. Price's seems to have been enlarged during the night. No other movements of the enemy are visible at this time.

Professor T.S.C. Lowe's Official Report

Respectfully, your obedient servant,
T. S. C. LOWE,
Chief Aeronaut, Army of the Potomac.

BALLOON CAMP,
Near Doctor Gaines', June 19, 1862--5.30 a.m.

Brigadier-General MARCY,
Chief of Staff:

GENERAL: I ascended at 4.30 o'clock this a.m. and remained up until after 5 o'clock, when the enemy's smokes became so numerous on our left that small objects, earth-works, &c., could no longer be defined. The enemy still have artillery near James Garnett's house, and their pickets on the side of the field toward Fair Oaks extend along the edge of the field near the woods.

The enemy appears not to be half so numerous on our right, and at this hour there are no movements of troops or wagons (save a few scattering ones) upon any of the visible roads.

T. S. C. LOWE.

P. S.--Preparations are going on to inflate a balloon near headquarters, which I hope to have ready to-day.

LOWE.

The principal observations being taken near headquarters, verbal reports were generally made, and I have no copies of any from the 19th to the 27th of June.

On the 26th I reported verbally to General Humphreys that the enemy had crossed the Chickahominy in large force, and was

Professor T.S.C. Lowe's Official Report

engaging our right wing at Mechanicsville. At daybreak next morning I received the following order:

FRIDAY, *June* 27, 1862.

Professor LOWE:

DEAR SIR: Ascensions must be made throughout the day, if practicable, at short intervals and reports made of what is seen.

A. A. HUMPHREYS.

JUNE 27, 1862---8.15 *a.m.*

The heaviest cannonading at this time is near where the last headquarters were, between Doctor Gaines' house and Mechanicsville. We have large reserves across the river; our forces are in line of battle. On our left the enemy appear to be in large force in and about their intrenchments on this side of the river in the vicinity of. Doctor Friend's, and on this side very large.

The dense smoke prevents me from seeing to Richmond. I am very unwell, and think it advisable for some good person to be constantly up.

Respectfully,
T. S. C. LOWE.

JUNE 27, 1862--9.20 *a.m.*

Brigadier-General HUMPHREYS, *or*
General *MARCY,*
Chief of Staff:

Professor T.S.C. Lowe's Official Report

Although I reported myself ill on this occasion I will remain constantly in the balloon, and if you will send me two orderlies I will keep headquarters constantly informed of what can be seen from the balloon. My assistants that you speak of are trying to save the property in their charge. In an exact north direction from here, and about two miles and a half from the river, in an open field, there are large bodies of troops, but I should judge they were too far down on our right to be the enemy. On a hill this side of Doctor Gaines' house there is a long line of skirmishers stationary. On the field near where General Morell was camped everything is on fire.

About four miles to the west from here the enemy have a balloon about 300 feet in the air. By appearances I should judge that the enemy might make an attack on our left at any moment. We are firing occasional shots on our left.

T. S. C. LOWE.

JUNE 27, 1862--11 a.m.

Brigadier-General HUMPHREYS, *or*
General MARCY,
Chief of Staff:

There is no firing on either side at this time. In a northerly direction, and about three or four miles from Woodbury's Bridge, there is a long line of dust running toward the York River Railroad. Quite a large body of the enemy are visible in the field where General Smith was camped, near the old headquarters. The rebel balloon suddenly disappeared about one hour since.

The enemy in front of here remain silent in and around their earth-works and rifle-pits.

T. S. C. LOWE.

P. S.--Can Major Webb come over and ascend?

Professor T.S.C. Lowe's Official Report

T. S. C. L.

Other reports were made at short intervals during the rest of the day, and at 6 o'clock I reported that the enemy on Gaines' Hill were making a desperate advance, while a large column was moving to outflank our forces on the extreme right, and evidently intended to intercept our crossing at Woodbury's Bridge. Soon after this report was made our reserves were sent to protect the crossing and to relieve those troops who had been engaged for two days.

I have no doubt that the information given in the above reports (from what I saw myself and have since learned) saved a large portion of our troops then engaged from being taken prisoners, and also caused a strong guard to be placed at Bottom's Bridge and other crossings below, which prevented the enemy from getting into our rear.

On the evening of the 28th I received orders to pack up everything pertaining to the aeronautic department and to be ready to move. Owing to the want of transportation to carry material for gas, the balloons were not put in use again until we reached Harrison's Landing. Here I was taken very ill with fever, which had been gradually coming on me for two or three weeks, and I was compelled to leave the army, placing the management of the aeronautic operations in charge of Mr. C. Lowe, who kept the balloon in use during the time the army remained at that place. On one occasion Commodore Wilkes had the balloon taken on the river, and while at an elevation of 1,000 feet was towed by a steamer, while the banks and country for miles back were examined.

The following order was received from General Humphreys relative to moving from Harrison's Landing:

AUGUST 13, 1862.

Mr. LOWE:

Professor T.S.C. Lowe's Official Report

DEAR SIR: The balloon department will, as far as possible, go by water in the Rotary. The barge or flat will be taken also. They will keep near or accompany the steamer carrying the surplus baggage to headquarters. Colonel Ingalls will inform you which that is. The details for the balloon department will march under the orders of the officer commanding them. They will take not less than six days' rations. The wagons, teams, &c., will be turned over to the quartermaster's department.

Perhaps one wagon may be retained to accompany the detachment of enlisted men.

A. A. HUMPHREYS.

All transportation, &c., now being in the hands of the quartermaster's department, it was necessary for me to have an order from the commanding general before I could reorganize the aeronautic department. On the arrival of the Army of the Potomac from the Peninsula I therefore addressed the following note to Colonel Colburn, assistant adjutant-general:

NATIONAL HOTEL,
Washington, D.C., September 5, 1862.

Col. A. V. COLBURN,
Assistant Adjutant-General:

COLONEL: Having recovered from my late illness, I came to Washington several days since hoping that I might be of service on the present occasion. I beg of you to remind the general that I am anxiously awaiting orders, and, as ever, ready and willing to serve him. Some balloon observations at this time might be of great advantage. I have everything ready to operate at a moment's notice.

Very respectfully, your obedient servant,
T. S. C. LOWE,
Aeronaut.

Professor T.S.C. Lowe's Official Report

I was answered by Colonel Colburn that my services would probably soon be required, but to remain in Washington until I received orders, as the general did not yet know when he would want to use the balloons.

I received no orders until the morning after the battle of Antietam, when a dispatch came from General Marcy to come to Sharpsburg with the balloons without delay. I started immediately, and on the third day from Washington I arrived with the train at Sharpsburg. The delay was occasioned by General A. A. Humphreys being ordered to take command of a division, and the aeronautic department having been left without the proper authority being vested in me to act independently, I was unable to accompany the army as formerly.

During the battle of Antietam General McClellan remarked on several occasions that the balloon would be invaluable to him, and he repeated this to me when I arrived, assuring me that better facilities should be afforded me in future. It was evident that he was extremely anxious to obtain information of movements at certain points which could be furnished only by the aeronaut, which if he had obtained might have resulted in the complete defeat and utter rout of the enemy while trying to effect his escape across the Potomac. On this occasion he greatly felt the need of reports from the balloons, which, having been on so many previous occasions furnished without even being called for, were perhaps not sufficiently valued.

On the night of my arrival the balloons were made ready, and the next morning I pointed out the enemy, who were in force near Martinsburg, Va. The balloons were kept in use at this point until the rebel army left for Winchester, and one was also employed at Bolivar Heights. The observations made here in the vicinity of mountains 1,200 feet high, were mainly of use in enabling us to change our position and approach nearer to the enemy.

When the army took up its march into Virginia it moved in roads commanded by the mountains, and as it was not thought that balloon observations along this route were needed, I was ordered to proceed to Washington, to move out on the railroad, where

Professor T.S.C. Lowe's Official Report

better facilities for transportation, &c., could be had.
On the 1st of November I received the following:

HEADQUARTERS ARMY OF THE POTOMAC,
November 1, 1862.

Professor LOWE:

Under all the circumstances General McClellan thinks it best that you should return to Washington with everything pertaining to the balloon department, and hold that department in readiness to take the field at any very short notice. Acknowledge this.

S. WILLIAMS,
Assistant Adjutant-General.

There seemed to be no further use for balloons now until the army reached Fredericksburg.
In order that the new commander of the Army of the Potomac, General Burnside, might know that I was ready for duty, I addressed the following communication to his chief of staff:

HEADQUARTERS AERONAUTIC CORPS,
Washington, November 20, 1862.

Major-General PARKE,
Chief of Star, &c.:

GENERAL: Considering it necessary that the commanding general should be informed in relation to my operations, and the service that I am prepared to render, I would respectfully submit the following succinct statement:
First. The U.S. aeronautic department under my direction is in excellent condition, with all the improvements just added that over a year's continual operations and experience could suggest. I have at this time six superior silk balloons with portable gas-

Professor T.S.C. Lowe's Official Report

generating apparatus, which enables me to inflate a balloon at any point in three hours sufficiently to raise two men and ropes to an elevation of 1,000 feet or more. The balloons can be used with nearly, if not quite, as good success in winter as in summer.

Second. In order to facilitate my operations and making prompt reports, I was permitted by General McClellan to add for my use a telegraph train, with five miles of insulated wire, which will enable me to make reports directly from the car of the balloon while viewing the enemy's position. The line can be otherwise useful for transmitting other messages not connected with my department.

Third. It being often necessary to inflate a balloon at night, and having many times performed the same under difficulties, owing to the want of light, I have introduced a powerful oxyhydrogen or calcium light for that purpose. Aside from the benefits of this light for the above purpose, it can be used to great advantage for many other purposes where night-work is to be performed, such as felling timber, building bridges, crossing streams, building earth-works, &c. One of these lights would be sufficient for at least 2,000 persons to work by with as much convenience as by daylight, and the rays can be entirely hidden from any point where it is not desirable to show them. With this apparatus light can be thrown two miles distant sufficiently powerful to work by. The cost is trifling.

Fourth. I also have with me a set of powerful magnifying lenses with which a photograph of three inches square can be magnified to the size of twenty feet square. Thus it will be seen that a view taken at a distance too far for the objects to be discernible with the naked eye, could be easily distinguished with the magnifier. A map photographed and thus magnified would be found much easier to consult.

Fifth. I keep with my corps a large number of small signal balloons which can be used day or night. Fires of red, white, blue, or green can be attached, which will burn more than ten times as long as a rocket, and with much greater brilliancy, and therefore can be seen with more certainty, and costs no more for them than for rockets.

Professor T.S.C. Lowe's Official Report

Having reduced all of the above-mentioned branches to a practical everyday working, I can be called upon for any or all of them at any time without inconvenience to the main balloon operations, and with but little expense, as the same portable gas-works can be used for them all.

Not considering it necessary to give a detailed account of what may be done, but hoping soon to be called into active service again,

I remain, with great respect, your very obedient servant,
T. S. C. LOWE,
Chief of Aeronautics, &c.

On receipt of the above communication the following order was returned:

HEADQUARTERS ARMY OF THE POTOMAC,
Opposite Fredericksburg, November 24, 1862.

Professor LOWE:

The commanding general desires that you proceed to Washington and bring up the apparatus and material, so that an ascension can be made at this point as early as possible. He desires that the Quartermaster's Department furnish you such aid and assistance in Washington and *en route* that you may require.

Very respectfully, your obedient servant,
JNO. G. PARKE,
Chief of Staff.

The next day everything was moved down to the army, but as General Burnside had deferred his operations, he desired the balloon should not be shown to the enemy till he was ready to cross the river. On the 12th of December I received orders to get the balloon ready, and the following morning (being the day of the

Professor T.S.C. Lowe's Official Report

battle of Fredericksburg) ascensions were commenced, and during the day many staff officers ascended, and much valuable information was furnished the commanding general, whose headquarters being directly under the balloon, verbal communications only were given, and no written reports are therefore inserted. Several shots were fired at the balloon during the day, one striking about two miles beyond the balloon, passing close to it, and going in all about three miles and three-quarters from where it was fired.

Nearly all of my reports during the following month were given verbally.

The following report was forwarded on December 22, which shows the duty that the balloon was required to do while the army was lying still:

HEADQUARTERS AERONAUTIC CORPS,
December 22, 1862.

Major-General PARKE,
Chief of Staff:

GENERAL: By observations taken from the balloon to-day the enemy's position was very clearly defined. Their main camps are opposite to our left, and extend down the river from four to six miles, and three miles back. Earth-works appear to be thrown up on the next range of hills beyond the first line of woods, but nothing definite could be ascertained concerning them owing to the heavy smokes.

By moving a balloon farther down the river more information can be obtained. They do not appear to have withdrawn any of their forces.

Very respectfully,
T. S. C. LOWE.

CAMP NEAR HEADQUARTERS ARMY OF THE POTOMAC,
January 13, 1863.

Professor T.S.C. Lowe's Official Report

Major-General PARKE,
Chief of Staff, &c.:

GENERAL: Please find inclosed a copy of a lithograph representing the balloon signals. Should these signals meet with the further approval of the general commanding I would respectfully ask that I may be notified as early as possible that I may have prepared a sufficient number to operate successfully. I would recommend about thirty of each denomination.

Very respectfully, your obedient servant,
T. S. C. LOWE,
Chief of Aeronautics, &c.

The signals above alluded to are not intended to take the place of anything now in use, but are simply an addition to be used in case of emergency, where it was necessary to communicate a long distance. Further mention of this will be made hereafter.

The following orders and reports up to March 21 will be sufficient to show the principal duties performed by the aeronautic department:

HEADQUARTERS AERONAUTIC DEPARTMENT,
February 4, 1863.

General BUTTERFIELD,
Chief of Staff:

SIR: From an observation taken this afternoon the enemy appear still in camp about three miles west of Fredericksburg; also a large camp south by west, about eight miles. The largest camp noticed appears to be south from the city about fifteen miles; also a smaller camp east by south.

The balloons are constantly in readiness, and observations can be taken at any time when the weather will permit.

Professor T.S.C. Lowe's Official Report

Very respectfully, your obedient servant,
T. S. C. LOWE,
Chief of Aeronautics, Army of the Potomac.

HEADQUARTERS AERONAUTIC DEPARTMENT,
Camp near Falmouth, February 7, 1863.

General BUTTERFIELD,
Chief of Staff, Army of the Potomac:

SIR: According to your order I have taken advantage of all suitable weather for several days past to reconnoiter the enemy's position from the balloon. Yesterday in the afternoon the atmosphere was very clear, and from observations taken then and again to-day the various positions of the enemy could be determined by their camps and smokes. The line of hills opposite Fredericksburg and above and below the city appear to be occupied by a small force, divided into small squads, while the heaviest camp appears to be at or near Bowling Green.

Still farther beyond, say twenty-five miles from Fredericksburg, are heavy camp smokes, which I should judge was at the junction of the Virginia Central and Richmond and Fredericksburg Railroads. Off to the right of the city, about ten or twelve miles, and some distance back from the river, are quite large camp smokes (I should think that this camp was at Spotsylvania Court-House), while in a direct line from these and near the river appears to be a camp of much smaller size.

Very respectfully, your most obedient servant,
T. S.C. LOWE,
Chief of Aeronautics, Army of the Potomac.

Professor T.S.C. Lowe's Official Report

FEBRUARY 7, 1863.

T. S. C. LOWE,
Chief Aeronaut, &c.:

Your interesting report just received. What do you consider a large camp as mentioned in your report, and what a small one? About how many men?

Keep your balloon up all you can, and confine the knowledge gained to your reports to these headquarters.

Should like to have you locate camps on maps which General Warren will furnish you for the purpose.

DANL. BUTTERFIELD,
Major-General and Chief of Staff.

HEADQUARTERS AERONAUTIC DEPARTMENT,
February 23, 1863.

Major-General BUTTERFIELD,
Chief of Staff, Army of the Potomac:

SIR: I ascended with the balloon this p.m., but was unable to discover any change in the position of the enemy as far as I could see.

To the south and southeast the atmosphere was too smoky to enable me to see anything in relation to their camp. I will ascend again as soon as the atmosphere becomes clear and furnish you with a fuller report.

Very respectfully, your most obedient servant,
T. S. C. LOWE,
Chief of Aeronautics, Army of the Potomac.

Professor T.S.C. Lowe's Official Report

HEADQUARTERS ARMY OF THE POTOMAC,
February 24, 1863.

Professor LOWE:

SIR: The balloon ascension to be made between daylight and sunrise to-morrow a.m. should be made with a view to giving us most careful and accurate information as to the number of the enemy and their camps. Rumors that a large portion of their force had gone make it very desirable. You may be able to gain much credit for your branch of science by the care and accuracy and promptness of your report. Can't you take Lieutenant Comstock up with you?

Yours,
DANL. BUTTERFIELD,
Major-General and Chief of Staff.

HEADQUARTERS ARMY OF THE POTOMAC,
February 27, 1863.

Professor LOWE,
Balloon Corps:

SIR: I am requested by Major-General Butterfield to direct that you place a balloon at the disposal of Lieutenant Comstock, chief engineer.

Very respectfully,
WM. L. CANDLER,
Captain and Aide-de-Camp.

HEADQUARTERS ARMY OF THE POTOMAC,
Camp near Falmouth, Va., March 1, 1863.

Professor T.S.C. Lowe's Official Report

COMMANDING OFFICER SIXTH CORPS:

SIR: The commanding general directs that upon the application of Professor Lowe, balloonist, you furnish him with a detail of one officer, one sergeant, and thirty-five men to assist him in making an ascension near White Oak Church.

Very respectfully, your obedient servant,
S. WILLIAMS,
Assistant Adjutant-General.

HEADQUARTERS ARMY OF THE POTOMAC,
March 12, 1863.

Professor LOWE,
Chief of Balloon Corps:

PROFESSOR: The commanding general directs that you make frequent ascensions during the day, moving your balloon from right to left near the river. He desires that you make very close observations of the enemy, noticing any movements or work going on or changes made. Watch and note very carefully the fords and all along the river bank. Report promptly anything you may see.

Very respectfully, your obedient servant,
S. WILLIAMS,
Assistant Adjutant-General.

HEADQUARTERS AERONAUTIC CORPS,
March 12, 1863.

Major-General BUTTERFIELD,
Chief of Staff, Army of the Potomac:

Professor T.S.C. Lowe's Official Report

GENERAL: I have just received an order from the general-in-chief, through General Williams, directing me to make frequent ascensions, &c., which I have made preparation to do at every favorable moment.

I ascended early this morning from a point near Falmouth, but was unable to discover any movements of the enemy on the roads or near any of the visible fords. All the camps around Fredericksburg remain quiet as usual.

At about 8 o'clock I discerned working parties throwing up earth a short distance to the right of the city on the low land; also in the woods on the first ridge. I then moved the balloon some three miles up the river, where I can get a fine view as soon as the high wind now prevailing ceases.

I have just received a report from one of my assistants, who ascended with the balloon down the river at 6 o'clock this morning (by my direction). Up to 8 o'clock all was quiet on the left, or as far down as the aeronaut could see, and all the camps remained as usual.

Very respectfully, your most obedient servant,
T. S. C. LOWE,
Chief of Aeronautics, Army of the Potomac.

HEADQUARTERS AERONAUTIC DEPARTMENT,
Near Falmouth, March 13, 1863.

Major-General BUTTERFIELD,
Chief of Staff, Army of the Potomac:

GENERAL: Between 5 and 6.30 o'clock this morning both balloons ascended, one near White Oak Church and the other about three miles up the river. No movement of the enemy was visible at that time, but all appeared to be quietly in camp, as the smoke ascended from them all. The camp smokes at Bowling Green were distinctly seen, as also one near Scott's Dam, on Golin

Professor T.S.C. Lowe's Official Report

Run, of considerable size. There is also a camp and quite a number of tents opposite Taylor's Dam. The enemy are still throwing up earth a short distance to-the right of Fredericksburg with embrasures for field pieces.

Since early this morning the weather has been too squally to admit of ascending with the balloon. Every opportunity, however, shall be improved and reports made.

                Very respectfully, your most obedient servant,
                              T. S. C. LOWE,
                              Chief of Aeronautics.

            HEADQUARTERS ARMY OF THE POTOMAC,
                                March 17, 1863.

Professor LOWE,
Balloon Department:

PROFESSOR: The major-general commanding directs that you make an ascension, if your balloon is in readiness, immediately after dusk, or as soon as rockets with their colors and fires are visible; that you report the color, &c., of rockets--if any can be seen--in a northwesterly or westerly direction. The colors expected are to represent signals as follows:

One signal, green; one signal, green and red; one signal, red and white; one signal, red and green; one signal, white and red. Answering signal from intermediate stations, green. Knowing what signals are expected, you can, perhaps, more readily and surely discern them. Report with care.

                Very respectfully, your obedient servant,
                              PAUL A. OLIVER,
                        Lieutenant and Aide-de-Camp.

Professor T.S.C. Lowe's Official Report

HEADQUARTERS CAVALRY CORPS, *ARMY OF THE POTOMAC, March* 19, 1863.

General S. WILLIAMS,
Assistant Adjutant-General:

    Professor Lowe has an arrangement for transmitting information from distant points by signal balloons, which I think might be made available and valuable with cavalry operating in the field. I have thought the subject over a good deal, and if the professor can get authority to procure the necessary apparatus I will take measures to test and, if possible, put his plan in practice.

Very respectfully, &c.,
GEORGE STONEMAN,
Brigadier-General, Commanding Corps.

[Indorsement.]

HEADQUARTERS ARMY OF THE POTOMAC,
March 19, 1863.

    Respectfully referred to Professor Lowe, with the request that he will please state in substance the preparations the proposed plan will require and the probable expense of the same.

By command of Major-General Hooker:
S. WILLIAMS,
Assistant Adjutant-General.

HEADQUARTERS AERONAUTIC CORPS,
Camp near Falmouth, Va., March 20, 1863.

Professor T.S.C. Lowe's Official Report

Brig. Gen. S. WILLIAMS,
Assistant Adjutant-General, Army of the Potomac:

GENERAL: In answer to your inquiry concerning the preparation and probable expense of testing my plan for signals by balloons, I would respectfully state that the preparation will consist in getting the balloons made of the proper material and sizes with proper attachments; constructing a variety of characters to be attached to the balloons for day signals; arranging a variety of different colored lights of great power and brilliancy in order that they may be seen a great distance. The time required to get everything ready, I think, would be about one week. The arrangement once completed, any person of ordinary intelligence can use the signals. The cost of thoroughly testing will not exceed $300, after which, if brought into use, the cost of each balloon for conveying signals will not exceed $6, where a quantity is ordered at one time.

I remain, general, very respectfully, your most obedient servant,
T. S. C. LOWE,
Chief of Aeronautics, Army of the Potomac.

[Indorsement.]

HEADQUARTERS ARMY OF THE POTOMAC,
March 20, 1863.

Respectfully returned.

It was inferred the tests made proved the expediency and capacity of the plan. Has not Professor Lowe balloons and signals enough on hand of the kind proposed to show their merits for this purpose? If he has, a board will be ordered immediately to report upon them. Return these papers without delay.

Professor T.S.C. Lowe's Official Report

By command of Major-General Hooker:
S. WILLIAMS,
Assistant Adjutant-General.

HEADQUARTERS AERONAUTIC CORPS,
Camp near Falmouth, Va., March 21, 1863.

Brig. Gen. S. WILLIAMS,
Assistant Adjutant-General, Army of the Potomac:

GENERAL: In answer to your indorsement upon my communication of yesterday, I would respectfully say that I have not on hand any signal balloons of the size or quality sufficient to show the merit, or to carry up sufficient weight of material for which they are designed.

I have some few balloons left of those ordered by Major-General Burnside, for experiments, but were gotten up in a hurry, and made of very poor material, but the best that could be obtained at the time. They will do very well to use for instruction. I have on hand a quantity of colored fires, but will require to be arranged differently, with some addition, in order to give the full effect and brilliancy desired. I have not any of the proper material on hand for the flags. My extreme estimate of the expense of these experiments was based upon the supposition that a large number of the signals would require to be sent up, embracing every variety of lights, flags, and characters upon the balloons, in order to choose the most desirable.

I am, general, with great respect, your most obedient servant,
T. S. C. LOWE,
Chief of Aeronautics, Army of the Potomac.

[Indorsement.]

Professor T.S.C. Lowe's Official Report

HEADQUARTERS ARMY OF THE POTOMAC,
March 21, 1863.

Respectfully returned.
Under the circumstances not favorably considered. General Stoneman to be informed by Professor Lowe.

By command of Major-General Hooker:
S. WILLIAMS,
Assistant Adjutant-General.

By the decision in this matter General Stoneman was deprived of a very valuable means of communicating with the commanding general while operating in the interior of the enemy's country. With the signal balloons alluded to General Stoneman could have been heard from every night, and answered from Fredericksburg, which certainly in his last famous raid would have been of great value both to him and to General Hooker.

These intense lights by the aid of balloons, varying in size from ten to twenty feet in diameter, can be sent from 3,000 feet to three miles in the air, and can be seen from 15 to 100 miles, according to the size of the lights. At any rate I would not hesitate on any clear night (with the proper facilities) to guarantee to signal even to a greater distance.

*HEADQUARTERS ARMY OF THE POTOMAC,
March 21, 1863.*

Prof. T. S. C. LOWE, &c.:

By direction of the General-in-Chief, you will report on Monday morning next to the Committee of Congress on the Conduct of the War, now sitting in the Capitol.

Professor T.S.C. Lowe's Official Report

By command of Major-General Hooker:
S. WILLIAMS,
Assistant Adjutant-General.

BALLOON CAMP,
Near Falmouth, Va., March 22, 1863.

Professor LOWE:

SIR: Lieutenant Comstock went up to-day in the Washington. It was very calm, and I let the balloon ascend to an elevation of 2,000 feet, where he remained one hour and a half in full view of the enemy's camps and works for twenty miles distant. The balloon was then towed, at an elevation of 1,000 feet, three miles on our left, with him in the car of the balloon. He expressed himself gratified with the knowledge thus obtained.

Respectfully,
JAMES ALLEN,
Aeronaut, in Charge of Balloon Washington.

BALLOON CAMP,
Near Phillips' House, March 26, 1863.

Prof. T. S. C. LOWE:

SIR: Made an ascension this 12 m. The largest camps of the enemy that could be seen were south and southwest from Fredericksburg. One very extensive camp about eight miles south from the city. I also discovered what I judge to be earth-works (new) from four to six miles west of the city. If earth-works, they are extensive. Could discover nothing of note up the river.

Professor T.S.C. Lowe's Official Report

Yours, respectfully,
E. S. ALLEN,
Aeronaut.

MARCH 27, 1863.

Hon. B. F. WAVE,
Chairman of Committee on Conduct of the War:

    SIR: Please find accompanying this note fifty-one reports of observations taken by me from the balloons during the latter part of May and the month of June, 1862, and forwarded to headquarters Army of the Potomac. They embrace but a small portion of the observations taken, but are all of the copies that I can now readily reach. It will be found that some few of these reports are without date, which is accounted for from the fact that they were sometimes written while in the balloon car and sent down to be copied and forwarded, and the persons who did this neglected to place dates upon the copies retained, as they were not considered of further value.

I remain, with great respect, your most obedient servant,
T. S. C. LOWE,
Chief of Aeronautics.

BALLOON CAMP,
Near Falmouth, Va., March 27, 1863.

Professor LOWE:

    SIR: To-day the balloon Washington was taken six miles to the left, and Lieutenant Comstock, Colonel Upton, and Major -----

Professor T.S.C. Lowe's Official Report

-, ascended separately, all of whom spoke in the highest terms of the advantage of this movable observatory, after which she was taken to her moorings.

Respectfully.
JAMES ALLEN,
Aeronaut, in Charge of Balloon Washington.

HDQRS. AERONAUTIC DEPARTMENT, *ARMY OF THE POTOMAC,*
March 30, 1863.

Brig. Gen. S. WILLIAMS,
Assistant Adjutant-General, Army of the Potomac:

GENERAL: I herewith respectfully report myself returned for duty to the Army of the Potomac, having been relieved for the present from the duties for which I was ordered to report there on the 23d instant.

Very respectfully, your most obedient servant,
T. S. C. LOWE,
Chief of Aeronautics, Army of the Potomac.

The following report contains many interesting facts concerning the system of aeronautics now employed and others proposed, to which I would call special attention; also to a letter following of April 1 from the present aeronauts in the Army of the Potomac:

HEADQUARTERS AERONAUTIC DEPARTMENT,
Camp near Falmouth, Va., March 30, 1863.

Professor T.S.C. Lowe's Official Report

Brig. Gen. S. WILLIAMS,
Assistant Adjutant-General, Army of the Potomac:

GENERAL: On the 21st of this month I received from you an article setting forth a new plan for operating balloons for military purposes, proposed by a Mr. B. Englend, and referred to me for an expression of opinion and report. In consequence, however, of my time being occupied during the past week in Washington before the Committee of Congress on the Conduct of the War, I have not been able to make a report until now.

In examining the papers I find many misstatements concerning the present balloon operations, which, in justice to myself and those connected with this department, I feel in duty bound to set right.

First, then, in comparing the two methods, he states that "the time required to inflate a balloon by the present mode is fifteen hours," when in fact it never required over three hours and fifteen minutes, and since adding my last improvements Mr. Allen, one of my assistants, informs me that the gas now makes in two hours and thirty minutes instead of fifteen hours as represented.

Second. He states that the cost of inflating now for a simple inflation is $400, when the actual cost is only about $60 now; and when the iron (which we now obtain free of cost at the Washington Navy-Yard) had to be purchased, the cost was then in the neighborhood of $75, which, when divided into fourteen (the number of days the balloons will retain their power, on the average), the cost per day for gas will be about $5.30. Of course this does not include contingent expenses.

Third. Mr. Englend states that it now requires 12,000 pounds of acid and iron for a single inflation, when, in fact, that amount will keep two balloons inflated from three to four weeks.

Fourth. He states that it now requires twelve or fourteen wagons, when the facts are that it never did require over seven wagons to haul four balloons and appendages and material to keep them inflated, and all camp and garrison equipage for the whole aeronautic corps.

Now that I have made the above corrections, I will give my

Professor T.S.C. Lowe's Official Report

opinion (as I am ordered to do so) of the relative advantages between the method proposed and the one now employed.

First. According to the statement of Mr. Englend, it requires a bulk of 68,000 cubic feet to lift the same weight that now requires 15,000 cubic feet, much lees than a quarter of the capacity of the balloon which he proposes. After figuring the weight of the appendages, which he puts down at 750 pounds, he then has left 250 pounds ascensive power. Now, considering that nine-tenths of the ascensions now made require an ascensive power of 400 to 600 pounds in order to counteract the force of the wind against the side of a balloon, it is certain that with a bulk more than four times as large and weight and with less than a quarter of the power, it could not ascend at all; or, in other words, when the balloon of 15,000 cubic feet capacity lifting 1,000 pounds, with weight of apparatus and two persons, between 400 and 500 pounds, can ascend from 1,000 to 2,000 feet, the balloon of 68,000 feet capacity and weighing 750 pounds, with a lifting power of 1,000, could not be held by fifty men against the wind, and would be blown to the earth.

Second. I should say that it would be impossible to tow from place to place a balloon of the kind last mentioned; therefore should two ascensions be required at different points in one day (as is often the case, in order to make a full and correct report), the balloon would have to be inflated at each point, which would be another impossibility, and would involve the expense of $250, according to the cost set down for each inflation. Besides, the constant handling of the machinery must necessarily soon wear it out.

I would here take occasion to say that the balloons now in service have been in use for nearly two years; have been inflated from one to two months without changing the gas; have stood the storms of two winters, and are kept constantly ready to ascend at five minutes' notice (whenever the weather will admit), and ascend four times higher than ever was done (by ropes) before, These are circumstances which history affords no parallel in any country. Notwithstanding this, I would respectfully recommend that Mr. Englend be permitted to try his experiments in the field beside the

Professor T.S.C. Lowe's Official Report

present balloon operations, in order to compare fairly the relative advantages of the two upon precisely the same grounds that I was allowed to try my first experiments, namely, with his own balloon and apparatus and at his own expense.

In conclusion, I would beg to state that the knowledge I have acquired in the aeronautic art has cost me much means and expense and many years of hard labor; therefore I would most respectfully ask that this report will not be furnished to Mr. Englend or his associates, as I desire not to instruct any persons except in the U.S. service.

I remain, general, with great respect, your most obedient servant,
T. S. C. LOWE,
Chief of Aeronautics, Army of the Potomac.

BALLOON CAMP,
Near Falmouth, Va., April 1, 1863.

Prof. T. S. C. LOWE,
Chief of Aeronautics, Army of the Potomac:

SIR: In accordance with your request that I should furnish you with a report of my operations previous to my employment under your direction and my opinion of your system of aeronautics, that you may avail yourself of it in your report to the Secretary of War, I would most respectfully submit the following:

For a number of years previous to the breaking out of this war I followed the profession of an aeronaut, as then practiced by the leaders in that art. At the commencement of the rebellion I was induced by my friends to offer my services to the Government. I did so, and for the purpose of demonstrating what I could do I brought on two balloons in July, 1861. Some experiments were made before an officer of the Topographical Engineers, appointed for that purpose, After witnessing my operations he pronounced them unsatisfactory, although I had, as a general thing, been as

Professor T.S.C. Lowe's Official Report

successful as other aeronauts had previously been. After ascertaining what was expected of balloons, and under what circumstances they would have to be operated, in' order to meet the requirements of those not acquainted with the art, I came to the conclusion that balloons could not be introduced into the U.S. service without an entire different arrangement. Not only must decided improvements be made in the balloon and paraphernalia, but the balloon must be inflated at short notice, and at different points in the field, and for that purpose there was no apparatus yet invented. After thus summing up the matter I returned to my home in Providence and subsequently watched with much interest the report of your progress in aeronautics for war purposes, until in the spring of 1862 you invited me to join your corps, since which time I have received much valuable information and instruction from you in the use of your inventions, which now enables me to operate with entire success, and, I believe, satisfactory to you, as I have often had evidence.

In conclusion, I can conscientiously say that the Government is indebted to you alone for the introduction of this useful branch of the public service, and were it not for your improvements in the construction of balloons and invention of portable gas generators, your untiring perseverance, hard labor, and exposure, against great obstacles, aeronauts could never have been of service to our Army.

Balloons, as usually constructed, could not be kept inflated in heavy winds, and at best could not hold their power but a few hours, whereas now the balloons are kept constantly ready to go up, day or night. From their manner of construction and great strength they are able to withstand any storm, and enables the aeronaut to ascend in nearly all weathers, and are so impervious that they can be kept inflated for months with but little replenishing, and consequently trifling expense. These are qualities heretofore unknown in the history of aeronautics, and are merits that deserve the highest commendation.

I remain, professor, with great respect, your most obedient servant,
JAMES ALLEN,
Aeronaut.

Professor T.S.C. Lowe's Official Report

I cordially concur in the foregoing as regards the superiority of Professor Lowe's system of aeronautics over former attempts. I have been engaged in ballooning for a number of years past and have been employed under the direction of Professor Lowe for the past five months. I have received much valuable instruction from him in the use of his new system of aeronautics for army purposes, without which balloons could not be used to any advantage in the field.

E. S. ALLEN,
Aeronaut.

SPECIAL ORDERS No. 95.

HEADQUARTERS ARMY OF THE POTOMAC,
Camp near Falmouth, April 7, 1863.

* * * * * * * * * *

12. Capt. C. B. Comstock, Corps of Engineers, is assigned to the immediate charge of the balloon establishment, and hereafter no issues or expenditure will be made on account of the same, except upon requisitions and accounts approved by that officer.

By command of Major-General Hooker:
S. WILLIAMS,
Assistant Adjutant-General.

Professor LOWE.

ARMY OF THE POTOMAC,
Near Falmouth, April 9, 1863.

Professor T.S.C. Lowe's Official Report

Capt. C. B. COMSTOCK,
Corps of Engineers, Army of the Potomac:

CAPTAIN: I am notified by a copy of Special Orders, No. 95, of April 7, 1863, that the balloon establishment is placed in your charge. Will you therefore please inform me of what duties I am expected to perform under your direction, that I may know how to proceed without conflicting with your arrangements.

I remain, very respectfully, your obedient servant,
T. S. C. LOWE.

HEADQUARTERS ARMY OF THE POTOMAC,
April 10, 1863.

Hon. P. H. WATSON,
Assistant Secretary of War:

SIR: In view of the present situation of our forces in the vicinity of Charleston and Baton Rouge, I would respectfully beg leave to submit the following statement:

I have a faithful person (aeronaut) who has been operating under my direction in this department for over a year; therefore, inasmuch as I have another assistant and some soldiers whom I have instructed sufficiently to help manage the balloon here, Mr. Allen--the person alluded to--could be spared for one of the other places. A complete set of apparatus is ready and can be shipped at short notice if required. The balloons here are constantly ready, and are used nearly every day more or less, and I have made preparation to render the utmost service at the next battle. The report that you requested from me is in progress and will soon be completed. It required more time than I at first supposed.

Professor T.S.C. Lowe's Official Report

I remain, with great respect, your most obedient servant,
T. S. C. LOWE,
Chief of Aeronautics, Army of the Potomac.

CAMP NEAR FALMOUTH, *VA., April* 12, 1863.

Capt. C. B. COMSTOCK,
Corps of Engineers:

CAPTAIN: Between 5 and 7 o'clock this p.m. I made two ascents with the balloon near White Oak Church, and obtained a very good view of the enemy's camps for a distance of about five miles. Beyond that distance the atmosphere was quite smoky. Along the ridge for a distance of about seven miles the enemy's camps are quite numerous, the heaviest being southwest, south, and southeast from where the balloon is anchored. From appearances I should judge they are fully as strong as ever. A clearer atmosphere, however, will enable me to form a better idea of their relative strength, &c.

Very respectfully, your obedient servant,
T. S. C. LOWE.

On the 12th of April I received the following order and instructions, which, considering the services I had rendered for two years and the experience I had acquired, I respectfully submit to the Honorable Secretary were as unnecessary as they were unexpected. I would call especial attention to the following communications up to May 7, 1863 (at which time I left the Army of the Potomac), that the Honorable Secretary may judge of my conduct under very embarrassing circumstances:

HEADQUARTERS ARMY OF THE POTOMAC,
April 12, 1863,

Professor T.S.C. Lowe's Official Report

Mr. T. S. C. LOWE, &c.:

As I informed you yesterday, I do not think the interests of the public service require the employment of C. Lowe, your father, or of John O'Donnell. Please inform me whether you have, as desired, notified them of the fact.

I also stated to you that it might be necessary for the public interest to reduce your pay from $10 to $6 per day. I also mentioned some general rules to be observed by all civil employés connected with the balloons. Some of them are repeated here, and you will please notify your subordinates of them: No absences from duty without my permission will be allowed, and pay will be stopped for the time of absence.

In camp, when the wind is still, ascensions should be made at morning, noon, and night, the labor being equally divided among the aeronauts, and reports made to me in writing of all that is observed during the day. If anything important was observed it should be reported at once. These reports should give the bearings of the important camps observed, and the camps should be numbered from right to left, No. 1 being on the right. You, as having larger experience, are expected to make these ascensions frequently, and to be responsible that no camp disappears and no new one appears without its being reported at once. You will also be held responsible that the apparatus is kept in good order; that the aeronauts attend to their duty; that the necessary requisitions are sent in for supplies, and generally for the efficiency and usefulness of the establishment, as well as its economical management.

Very respectfully,
C. B. COMSTOCK,
Captain of Engineers and Chief Engineer Army of the Potomac.

I asked you yesterday for an inventory of all public property under your charge. Please send it to me to-morrow.

Professor T.S.C. Lowe's Official Report

CAMP NEAR FALMOUTH, *VA., April* 12, 1863.

Maj. Gen. D. BUTTERFIELD,
Chief of Staff, Army of the Potomac:

GENERAL: From a copy of Special Orders, No. 95, April 7, 1863, I am informed that the balloon establishment is placed in charge of Capt. C. B. Comstock, Corps of Engineers, to whom I reported immediately on receipt of the above order. In conversation with him yesterday I learned that different arrangements were to be made, and among other things he informed me that my compensation for services were reduced from $10 per day to $6. This Captain Comstock does, I have no doubt, in good faith, and from the view which he takes of this department as it now stands.

Now, in justice to myself and the service in which I am engaged, I beg to submit the following succinct statement:

At the breaking out of the rebellion I was urged to offer my services to the Government as an aeronaut. I did so, at the sacrifice of my long-cherished enterprise in which I had expended large sums of money and many years hard labor, and which, if successful, would compensate me for my expenditure and place aeronautics among the first branches of useful science.

(The enterprise above alluded to could not now be revived, except under the most favorable circumstances.)

During my first operations for the Government I had three competitors in the field and many more applicants. I used my own machinery and expended considerable private means, and two months' labor, for all of which I have never received pay.

My system of aeronautics was selected, and I was offered $30 per day for each day I would keep one balloon inflated in the field ready for officers to ascend. (This was when it was supposed balloons could not be kept constantly inflated, as is now the case.) I declined this offer and offered my services for $10 per day, as I desired to continue during the war and add to my reputation; besides, that amount would be sufficient to support my family. Ever since then I have labored incessantly for the interest of the

Professor T.S.C. Lowe's Official Report

Government, and I have never shrunk from duty or danger whenever it was necessary to gain information for the commanding general.

For nearly two years, aside from doing all the business of this department, I have made frequent personal reconnaissances and have attended to the management of several balloons for different officers to ascend until within the past two or three weeks, during which time I have been occupied by order of the Secretary of War in preparing a history of this branch of the service, &c., at the same time keeping an eye to the proper management of the balloons, which have been kept in constant use, attended by my assistants.

General, I feel aggrieved that my services should not have been better appreciated. As it is, I cannot honorably serve for the sum named by Captain Comstock without first refunding to the Government the excess of that amount which I have been receiving ever since I have been in the service. This my very limited means will not allow, for it requires full the salary I have received to support myself in the field and my family at home; therefore, out of respect to myself and the duty I owe my family, it will be impossible for me to serve upon any other conditions than those with which I entered the service.

Notwithstanding, as I have promised the commanding general that nothing should be lacking on my part to render the greatest possible service during the next battle, and as I consider that all should be done that genius can devise to make the first move successful, I will offer my services until that time free of charge to the Government.

> I remain, general, with great respect,
> **T. S. C. LOWE,**
> *Aeronaut.*

The following are five [four] indorsements made upon the foregoing document:

Professor T.S.C. Lowe's Official Report

HEADQUARTERS ARMY OF THE POTOMAC,
April 13, 1863.

Respectfully returned to Professor Lowe, to be forwarded through the proper channel to Captain Comstock, chief of engineers.

By command of Major-General Hooker:
S. F. BARSTOW,
Assistant Adjutant-General.

CAMP NEAR FALMOUTH, *VA., April* 13, 1863.

Respectfully forwarded to Capt. C. B. Comstock, chief engineer, Army of the Potomac.

It was supposed that this was properly addressed, and I take pleasure in rectifying the mistake.

T. S.C. LOWE,
Aeronaut.

Respectfully forwarded.

It is believed that during the two years Mr. Lowe has been receiving $10 per day for his services he has been compensated for the sacrifices made, and that $6 per day is ample payment for the duties he has to perform at present.

C. B. COMSTOCK,
Captain of Engineers and Chief Engineer Army of the Potomac.

Professor T.S.C. Lowe's Official Report

HEADQUARTERS ARMY OF THE POTOMAC,
April 15, 1863.

Respectfully returned.
See indorsement of Captain Comstock, Engineer Department, in charge of balloons.

By command of Major-General Hooker:
S. WILLIAMS,
Assistant Adjutant-General.

WAR DEPARTMENT,
Washington City, April 13, 1863.

T. S. C. LOWE,
Chief of Aeronautics, Headquarters Army of the Potomac :

SIR: The Secretary of War directs me to acknowledge the receipt of your letter of the 10th instant stating that you can spare an experienced aeronaut, should his services be required in the vicinity of Charleston or Baton Rouge, and that a complete set of balloon apparatus is ready and can be shipped at short notice. In reply the Secretary directs me to instruct you to have all necessary preparations completed as soon as possible. You will advise this Department of the weight and bulk of the apparatus and supplies, and also when and from what point the aeronaut you recommend will be ready to start.

Very respectfully, your obedient servant,
P. H. WATSON,
Assistant Secretary of War.

CAMP NEAR FALMOUTH, *VA., April* 19, 1863.

Professor T.S.C. Lowe's Official Report

Respectfully referred to Gen. S. Williams, assistant adjutant-general.

The within has been complied with, and Mr. James Allen named as the person that could be spared, inasmuch as I have another person to take his place here, and he would be best suited for another point.

In my judgment the above arrangement will not in the least interfere with the successful operations of the balloons in this army. Therefore I would respectfully recommend that Mr. Allen be ordered to report for the above duty at once.

Very respectfully,
T. S. C. LOWE,
Chief of Aeronautics, Army of the Potomac.

HEADQUARTERS ARMY OF THE POTOMAC,
April 19, 1863.

The accompanying communication is respectfully returned to Professor Lowe, to be forwarded through Captain Comstock, engineer, who is in charge of the balloon department. The commanding general desires to be informed why the letter to the Secretary of War, to which the answer is in reply, was not transmitted through headquarters.

By command of Major-General Hooker:
S. WILLIAMS,
Assistant Adjutant-General.

CAMP NEAR FALMOUTH, *April* 20, 1863.

Professor T.S.C. Lowe's Official Report

Capt. C. B. COMSTOCK,
Chief of Engineers, Army of the Potomac:

CAPTAIN: According to your directions, I referred the inclosed letter from the Assistant Secretary of War to General Williams, who has returned it with the accompanying note.

In answer to the commanding general, why my letter to the Assistant Secretary of War was not transmitted through headquarters, I would respectfully state that I was not aware that it was customary to do so, and if in my zeal to render service to the Government I have overstepped the bounds prescribed by military law I can only say that it was unintentional.

I remain, very respectfully, your obedient servant,
T. S. C. LOWE,
Chief of Aeronautics, Army of the Potomac.

HEADQUARTERS ARMY OF THE POTOMAC,
April 20, 1863.

Respectfully forwarded, and indorsement of T. S. C. Lowe not approved.

C. B. COMSTOCK,
Captain of Engineers and Chief Engineer Army of the Potomac.

HEADQUARTERS ARMY OF THE POTOMAC,
April 20, 1863.

On the 19th instant Mr. T. S.C. Lowe, aeronaut, informed me that he had been directed by the Honorable Secretary of War to send a balloon and aeronaut to Charleston, and that he had selected

Professor T.S.C. Lowe's Official Report

Mr. J. Allen. At my request he showed me the accompanying letter from the Assistant Secretary of War.

I informed him that such orders should come to me from the adjutant-general of this army, and not from himself; that he, not being in charge of the balloon establishment, had not the power to change it; and that I did not think it consistent with the interests of this army to detach Mr. J. Allen from it at present. A balloon can be spared without detriment.

Respectfully forwarded to adjutant-general, Army of the Potomac.
C. B. COMSTOCK,
Captain of Engineers and Chief Engineer Army of the Potomac.

HEADQUARTERS ARMY OF THE POTOMAC,
April 21, 1863.

Respectfully returned.

Captain Comstock will make the necessary arrangements for the balloon to be placed at the disposal of the War Department and advise the Assistant Secretary of War, as herein directed.

If it is possible for him to spare an aeronaut he will name the one selected in his communication concerning the balloons.

By command of Major-General Hooker:
S. WILLIAMS,
Assistant Adjutant-General.

HEADQUARTERS ARMY OF THE POTOMAC,
Camp near Falmouth, Va., April 15, 1863.

Professor T.S.C. Lowe's Official Report

Hon. P. H. WATSON,
Assistant Secretary of War, Washington, D.C.:

    SIR: Your letter of the 13th instant is received, and in answer would respectfully state that the weight and bulk of the apparatus and supplies necessary for the balloon to be sent South or West are as follows. Two balloons and appendages, about 500 pounds, in a basket three feet by five and two feet deep.

    One set of gas generators to go in two army-wagon running gears, same dimension as wagon body and five feet high, weighing about 1,000 pounds each. Material to keep one balloon inflated day and night for two months will consist of 100 carboys of sulphuric acid, weighing about 16,000 pounds, and twenty barrels of iron turnings, weighing about 10,000 pounds. The cost of the above amount of gas material, as now purchased, is about $350--less than $6 dollars per day. The acid can be obtained from Messrs. Savage & Stewart, No. 18 North Front street, Philadelphia, Pa.; the iron at the Washington Navy-Yard. The aeronaut, Mr. James Allen, will be in Washington on Monday next, with everything complete and ready to start from that point, provided the quartermaster procures the acid and iron above mentioned. The salary required by Mr. Allen is $5 per day with rations, or $5.75 per day without rations, and all necessary transportation.

    I remain, sir, with great respect, your most obedient servant,
T. S. C. LOWE,
Chief of Aeronautics, Army of the Potomac.

BALLOON CAMP,
Near Falmouth, April 14, 1863.

Professor LOWE,
Chief of Aeronautics:

Professor T.S.C. Lowe's Official Report

An extensive camp seven miles southwest of Sherwood's forest; one extensive camp southeast of Sherwood's forest, about five miles; one southwest of the left of our picket line, about four miles from the river; one extensive camp eight miles from the left of our picket line in a south-southwesterly direction. About ten miles from Sherwood's forest in a westerly direction I saw a large column moving to our right, or the left of the enemy.

I am, sir, *yours,* respectfully,
JAMES ALLEN,
Aeronaut.

HEADQUARTERS AERONAUTIC DEPARTMENT,
Camp near Falmouth, Va., April 14, 1863.

Capt. C. B. COMSTOCK,
Chief Engineer, Army of the Potomac:

CAPTAIN: On hearing that Mr. Allen saw a column (while in the balloon near White Oak Church) moving to the right, I immediately went up in the balloon near Falmouth Station to observe if any extra camp smoke or fires could be seen to the west, but was unable to notice any change, except a few camp-fires not noticed before, on the road from Fredericksburg toward Chancellorsville, I should judge about six miles. All the rest of the camps remain the same as usual.

This p.m. three regiments were drilling on the fiats, two to the south and one to the right of Fredericksburg.

The following are the compass bearings of the various camps, as seen by Mr. E. S. Allen from balloon near Falmouth Station.

Extreme right to extreme left: No. 1, 8 to 4 miles west; No. 2, 2 miles west by south; No. 3, 2 miles southwest by west; No. 4, 2 to 3 miles southwest: No. 5, 2 to 3 miles southwest by south; No. 6, 2 miles south; No. 7, 4 to 5 miles south: No. 8, 8 to 10 miles south.

Professor T.S.C. Lowe's Official Report

The usual amount of smoke arose from all the above camps this evening.

It is evident, from all appearances, that the enemy have not made any considerable move as yet.

The balloons will be up at daybreak if the weather will admit.

Very respectfully,
T. S. C. LOWE,
Chief of Aeronautics, Army of the Potomac.

CAMP NEAR FALMOUTH, *VA., April* 17, 1863.

Capt. C. B. COMSTOCK,
Chief Engineer, Army of the Potomac:

CAPTAIN: During my observations to-day I was unable to discover any changes in the position of the enemy. The following is the compass bearing, taken of the enemy's position by Mr. Allen, from the Phillips house, which I find to be as near correct as is possible to get from that point.

Position of the enemy's camps as observed from balloon Eagle, April 17, 1863, beginning with the most distant one, west from Phillips' house, Va.:

No. 1, west 5 miles (large camp); No. 2, west by south 3 miles; No. 3, west by south 6 to 8 miles; No. 4, southwest by west 2 miles (large camp); No. 5, southwest by west 12 to 15 miles (large camp); No. 6, southwest 3 miles; No. 7, southwest by south 3 miles; No. 8, southwest by south 10 to 12 miles (large camp); No. 9, south 2 miles (large camp); No. 10, south 3 to 4 miles; No. 11, south 8 to 10 miles (large camp). Three or four small camps near the river bank, south by east.

Very respectfully,
T. S. C. LOWE,
Chief of Aeronautics.

Professor T.S.C. Lowe's Official Report

CAMP NEAR FALMOUTH, *VA., April* 18, 1863.

Capt. C. B. COMSTOCK,
Chief Engineer, Army of the Potomac:

CAPTAIN: Inclosed is Mr. Allen's report of observations taken to-day. I ascended this p.m. (the atmosphere being clearer in the west) and could see no change. The camp smoke arose from the usual places as far as I could see.

I could not get very high, however, in consequence of the strong breeze blowing at the time.

Very respectfully,
T. S. C. LOWE.

HEADQUARTERS ARMY OF THE POTOMAC,
April 19, 1863.

T. S. C. LOWE,
Chief Aeronaut:

Please inform me what has been the custom when on the march. Have the balloon guard moved with the balloon trains? And are two escorts, namely, the two details we now have needed, or only one, or none, in case of a movement?

Please let me know what material you think should go when we move.

These things should all be thought of and arranged, my approval only being needed.

Very respectfully,
C. B. COMSTOCK,
Captain of Engineers and Chief Engineer Army of the Potomac.

Professor T.S.C. Lowe's Official Report

CAMP NEAR FALMOUTH, *VA., April* 19, 1863.

Capt. C. B. COMSTOCK,
Chief of Engineers, Army of the Potomac:

    CAPTAIN: In answer to yours of this date asking what has been the custom when on the march, and whether the present escort are needed or not, I would state that it has been customary for the men detached on the balloon service to accompany the aeronautic train in order that balloon observations may be taken along the route when required.

    I would recommend that the details for both balloons be retained, inasmuch as considerable pains have been taken to instruct them in the requirements of the department. This will enable us to tow the balloons along as the army advances and take observations whenever required; and should bad weather compel us to discharge the gas, sufficient material should be taken along to reinflate, which can be done in the night, and observations taken of the enemy's position and the roads they take at daylight in the morning. I anticipate that the balloon can be of more service when moving than at any other time, provided we are following the enemy. I informed Captain Howard, assistant quartermaster, what transportation would be necessary for this department, and he informs me that he has set the same aside for our use, namely, seven wagons.

Very respectfully, yours, &c.,
T. S. C. LOWE,
Chief of Aeronautics.

HEADQUARTERS ARMY OF THE POTOMAC,
April 20, 1863.

Mr. T. S. C. LOWE,
Aeronaut:

Professor T.S.C. Lowe's Official Report

Please send me the names of three or four persons whom you deem best qualified to take charge of an independent balloon, with their addresses, not including those aeronauts With this army.

Respectfully,
C. B. COMSTOCK,
Captain, &c.

CAMP NEAR FALMOUTH, *April* 20, 1863.

Capt. C. B. COMSTOCK,
Chief Engineer, Army of the Potomac:

CAPTAIN: In answer to yours of this date asking for the names of three or four persons whom I deem best qualified to take charge of a balloon, I would respectfully say that I cannot name but two persons whom I could recommend for the Government service, aside from those already employed, although if occasion requires it I might select several who could be instructed in the use of army balloons.

The names of the two persons above alluded to are Mr. W. S. Morgan, No. 293 Second street, Jersey City, N.J., and Mr. J.B. Starkweather, Boston, Mass. Both of these parties, placed under an experienced army aeronaut, would render good service.

Very respectfully, &c.,
T. S. C. LOWE,
Chief of Aeronautics, Army of the Potomac.

CAMP NEAR FALMOUTH, *VA., April* 21, 1863.

Captain COMSTOCK:

Professor T.S.C. Lowe's Official Report

I ascended at about sundown this evening, but the atmosphere was too hazy to admit of a detailed examination of the enemy's position. All the principal camps, however, were visible and appear unchanged, I have taken a large balloon (capable of taking up two persons) to the left this p.m.

Respectfully, &c.,
T. S. C. LOWE,
Chief of Aeronautics Army of the Potomac.

HEADQUARTERS ARMY OF THE POTOMAC,
April 21, 1863.

Mr. T. S. C. LOWE,
Aeronaut:

Please have a balloon put in condition, so far as is practicable here, to be placed at the disposal of the Honorable Secretary of War at once. Please also inform me when it and machinery will be ready to be turned over to the quartermaster for transportation, and if there are any repairs needed which cannot be done here or anything needed to its efficiency not to be obtained here, please furnish me with a statement of such things in full. Also make out a list of everything needed to go with it. Also please inform me which of the two persons recommended by you as aeronauts a few days ago you deem best qualified to accompany the balloon.

Very respectfully,
C. B. COMSTOCK,
Captain of Engineers.

CAMP NEAR FALMOUTH, *VA., April* 21, 1863,

Professor T.S.C. Lowe's Official Report

Capt. C. B. COMSTOCK,
Chief of Engineers, Army of the Potomac:

    CAPTAIN: In answer to yours of this date I would respectfully say that all of the balloons, with the exception of the two now in use (needing repairs that could not readily be done in the field), were:sent to Washington on the 17th with the balloon barge and old generators, which also need repairs. The balloons were sent to the Columbian Armory, where they have always been taken for repairs or storage, there being a large room for that purpose.

    I intended four balloons to be kept in readiness for this army, and that two should be sent with the aeronaut that goes South, in order that he may operate with economy and to advantage. As to repairs to the balloons, it will be impossible to state exactly what they are until they are thoroughly examined. The principal things, however, for the two that I intended for the South are turning inside out, recoating, and inserting new top and valve in one of them.

    As to the two aeronauts, of whom you desire me to name the one best qualified to be placed at the disposal of the War Department, I would state that, in my opinion, for that service neither of them would answer, unless directed by an experienced army aeronaut who has had experience in the management of balloons for war purposes, which is quite different from the art practiced in the ordinary way. Therefore if you do not desire to send the aeronaut first named by me, under all the circumstances I would most respectfully ask to be ordered to report to the Secretary of War in his stead. With this arrangement the wishes of the Honorable Secretary could be complied with, and at the same time all machinery could be kept in order for all points where balloons are used.

    I remain, very respectfully, your obedient servant,
T. S. C. LOWE,
Aeronaut.

Professor T.S.C. Lowe's Official Report

CAMP NEAR FALMOUTH, *VA., April* 22, 1863.

Capt. C. B. COMSTOCK,
Chief of Engineers, Army of the Potomac:

CAPTAIN: I examined the enemy's position more closely this p.m., between 4 and 6 o'clock, than I have had an opportunity of doing for a number of days past. If I might be permitted to venture an opinion as to the relative strength of the enemy, I should say that they are about three to our four. I should estimate their supports to the batteries immediately back of the city of Fredericksburg to be about 10,000.

Immediately opposite where General Franklin crossed, say from two to three miles from the river, and from the railroad station along the height about one mile and a half, I should say that there were 25,000 troops camped.

Still farther to the left and south of the railroad there are also several large camps. During the time I was up I noticed many regiments on parade, near the various camps, and at one place there were three, while still farther back, I should judge four miles from the river and one mile to the left of the railroad, I saw a column of infantry moving to the right which required about twenty minutes to pass a given point, after I discovered them, and I counted what I took to be seven regiments. They had no colors flying as those did that were on parade.

Should the morning be fine I should be gratified to ascend with you, and could then better explain the points referred to. I am inclined to believe that the enemy are either strengthening their army or bringing up their troops from Bowling Green and the Junction. The latter is the most probable, as there is not as much smoke visible in that direction as heretofore.

Very respectfully, your obedient servant,
T. S. C. LOWE,
Chief of Aeronautics.

Professor T.S.C. Lowe's Official Report

On the 27th General Butterfield ordered me to make frequent ascensions, and to report to him and to General Sedgwick. Captain Comstock was then absent, and I did not see him until the 6th of May.

The following orders and reports relative to observations during the seven-days' battle I think worthy of special attention, as they show what can be done by the balloons when required, and they demonstrate their value as a means of observation, although there might be occasions when even more service could be rendered:

> HEADQUARTERS ARMY OF THE POTOMAC,
> *April 28, 1863.*

Professor LOWE,
Chief of Balloon Department:

PROFESSOR: The general commanding desires you to have your balloon up to-night, to see where the enemy's camp-fires are. Some one acquainted with the position and location of the ground and the enemy's forces should go up.

> Very respectfully,
> **PAUL A. OLIVER,**
> *Lieutenant and Aide-de-Camp.*

> BALLOON IN THE AIR,
> *April 29, 1863--10 a.m.*

Major-General SEDGWICK,
Commanding Left Wing, Army of the Potomac:

GENERAL: The enemy's line of battle is formed in the edge of the woods at the foot of the heights from opposite

Professor T.S.C. Lowe's Official Report

Fredericksburg to some distance to the left of our lower crossing. Their line appears quite thin compared with our force. Their tents all remain as heretofore, as far as can be seen.

T. S. C. LOWE,
Aeronaut.

12 M.

The enemy's infantry are moving to our right about four miles below our crossing on a road just beyond the heights. The enemy do not appear to advance.

T. S. C. LOWE.

1.30 P.M.

The enemy are moving wagon trains to their rear. Their force, which is in position opposite our crossing, is very light. I should judge not more than we now have across the river.

T. S. C. LOWE.

2.45 P.M.

About two regiments of the enemy's infantry have just moved forward from the heights and entered the rifle-pits opposite our lower crossing. Heavy smokes are visible about six miles up the river on the opposite side in the woods.

Professor T.S.C. Lowe's Official Report

T. S. C. LOWE,
Chief of Aeronautics.

HEADQUARTERS ARMY OF THE POTOMAC,
April 29, 1863.

Professor LOWE, *&c.:*

The major-general commanding directs that one of your balloons proceeds to-night or before daybreak to-morrow to Banks' Ford, or vicinity, and takes position to ascertain with regard to the force of the enemy between Fredericksburg, Bowling Green, and Banks' Ford. A signal telegraph is working between here and Banks' Ford, by which information can be communicated.

It is especially desired to know the comparative strength of the enemy's force at Franklin's Crossing, and in the vicinity of Banks' Ford, and above to the west of Fredericksburg.

BUTTERFIELD,
Major-General and Chief of Staff.

HEADQUARTERS ARMY OF THE POTOMAC,
April 29, 1863.

Maj. Gen. J. SEDGWICK,
Commanding Sixth Corps:

GENERAL: The commanding general desires that you will please have the accompanying communication sent at once to Professor Lowe, who is supposed to be in your vicinity.

Professor T.S.C. Lowe's Official Report

Very respectfully, &c.,
S. WILLIAMS,
Assistant Adjutant-General.

HEADQUARTERS ARMY OF THE POTOMAC,
April 29, 1863.

Professor LOWE, *&c.:*

The major-general commanding directs that your balloon on service near Sedgwick's command be sent up at a very early hour in the morning before sunrise, and that you get a communication with the signal telegraph to forward to these headquarters the earliest information with regard to the numbers, strength, and position of the enemy. This is not to interfere with the service of the balloon at Banks' Ford.

Very respectfully, your obedient servant,
S. WILLIAMS,
Assistant Adjutant-General.

APRIL 29, 1863

JAMES ALLEN,
In Charge of Balloon Washington:

You will have your men prepare one or two days' rations to-night, and in the morning have the men all ready to cross the river by daybreak. I will meet you where the balloon is now anchored.

Professor T.S.C. Lowe's Official Report

Very respectfully, &c.,
T. S. C. LOWE,
Chief of Aeronautics.

HEADQUARTERS AERONAUTIC CORPS,
April 29, 1863.

Mr. E. S. ALLEN,
In Charge of Balloon Eagle:

 General Hooker desires a reconnaissance made after dark to observe the location of the enemy's camp-fires. Also in the morning immediately before daybreak. Great care should be taken to gain all the information you can. Please make a careful report after 9 o'clock to-night and soon after daylight in the morning. A high altitude should be attained in order to accomplish the object desired. Be careful you observe the points of the compass correctly.

Very respectfully, your obedient servant,
T. S. C. LOWE,
Chief Aeronaut, Army of the Potomac.

HEADQUARTERS ARMY OF THE POTOMAC,
April 29, 1863.

Maj. Gen. JOHN SEDGWICK:

 GENERAL: I shall be absent to-morrow morning at Banks' Ford and vicinity, and if I may venture an opinion, I think it advisable that some engineer or other competent officer be instructed to ascend in balloon Washington from time to time until

Professor T.S.C. Lowe's Official Report

my return, for the purpose of reconnoitering from Fredericksburg as low down as the commanding general deems necessary.

       Very respectfully, your obedient servant,
           T. S. C. LOWE,
       Chief Aeronaut, Army of the Potomac.

      HEADQUARTERS ARMY OF THE POTOMAC,
         April 29, 1863--10 *p.m.*

Mr. E. S. ALLEN:

  The commanding general directs that your balloon be taken to Banks' Ford in order to take very important observations before and after daybreak. I will be there at daybreak, but you can commence to take observations should I not be there in time. The best way to go is to follow the signal telegraph. Look out for obstructions, &c., and don't fail, for now is your time to gain a position.

           Respectfully, &c.,
           T. S. C. LOWE,
     Chief of Aeronautics, Army of the Potomac.

     BANKS' FORD, *April* 30, 1863--10.45 *a.m.*

Maj. Gen. BUTTERFIELD,
 Chief of Staff, &c. :

  The balloon arrived at 3 a.m., but since that time have not been able to get an observation until now. The enemy opposite here are apparently not near as strong as they are opposite

Professor T.S.C. Lowe's Official Report

Franklin's Crossing, while opposite United States Ford there appears to be only one camp. I cannot yet see to Bowling Green, owing to the low clouds. The enemy's smokes are more numerous than usual in the rear of the heights opposite Franklin's Crossing below Fredericksburg.

T. S. C. LOWE,
Aeronaut.

BANKS' FORD, *April* 30, 1863--*l.*30 *p.m.*

Maj. Gen. BUTTERFIELD, *&C.:*

The enemy opposite this ford occupy three positions from a half to one mile from the river, also opposite what I take to be United States Ford. About five miles up there is a small force. To the left of Banks' Ford, commanding the road, the enemy have a battery in position. It is hard to estimate their force, for they are partially concealed in the pine woods, but they are certainly not near as strong as below Fredericksburg.

Respectfully, &c.,
T. S. C. LOWE.

4.45 P.M.

The enemy opposite this place remain the same as last reported. Numerous camp smokes are now arising from the woods, about ten or twelve miles in a southwest by westerly direction.

T. S. C. LOWE.

Professor T.S.C. Lowe's Official Report

HEADQUARTERS AERONAUTIC CORPS,
Camp near Falmouth, April 30, 1863---8.30 *p.m.*

Major General BUTTERFIELD,
Chief of Staff:

GENERAL: After my report at 4.45 this p.m. I came down to General Sedgwick's headquarters and ascended at 7 o'clock, remaining up until after dark in order to see the location of the enemy's camp-fires. I find them most numerous in a ravine about one mile beyond the heights opposite General Sedgwick's forces, extending from opposite the lower crossing to a little above the upper crossing. There are also many additional fires in the rear of Fredericksburg. From appearances I should judge that full three-fourths of the enemy's force is immediately back and below Fredericksburg.

Very respectfully, your most obedient servant,
T. S. C. LOWE,
Chief of Aeronautics, Army of the Potomac.

This last report was of much importance, as it gave the commanding general correct information as to the position of the enemy, and he was enabled to regulate his operations at Banks' and United States Fords accordingly. I was confident that the enemy had brought up reserves from Bowling Green and the Junction, and this induced me to hasten to Franklin's Crossing to take an observation there the same evening, although I was considerably exhausted from having been up all the previous day and night. I also concluded from General Hooker's movements that the enemy would learn them, and probably move up the river the next morning. I accordingly sent the following order to an assistant in charge of the balloon at Banks' Ford, and to this and the reports I made on the following morning I would call attention.

APRIL 30, 1863.

Professor T.S.C. Lowe's Official Report

Mr. E. S. ALLEN,
In Charge of Balloon Eagle, Banks' Ford:

Commence observations at daylight to-morrow morning, and look out for the enemy moving on the roads, either up or down, and report by telegraph, having your dispatch sent to General Hooker at United States Ford, and to General Sedgwick, Franklin's Crossing. Be sure of the correctness of your reports, and report promptly.

T. S. C. LOWE,
Chief of Aeronautics, Army of the Potomac.

The following eight dispatches were of the greatest importance, and especially when it is considered that all of these movements were out of sight of all but the observer in the balloon, and the information could not have been obtained in any other way:

BALLOON IN THE AIR,
May 1, 1863--9.15 a.m.

Major-General SEDGWICK,
Commanding Left Wing, Army of the Potomac:

GENERAL: Heavy columns of the enemy's infantry and artillery are now moving up the river accompanied by many army wagons, the foremost column being about opposite Falmouth and three miles from the river. There is also a heavy reserve on the heights opposite the upper crossing, and all the rifle-pits are well filled.

T. S. C. LOWE.

Professor T.S.C. Lowe's Official Report

9.30 A. M.

Still another column has just started from opposite the upper crossing, but not those mentioned as reserved in my last dispatch. They are moving with great rapidity.

T. S. C. LOWE.

10 A. M.

A column of the enemy are now crossing a small run that empties into the Rappahannock at Banks' Ford. One of the columns that left from opposite here required thirty minutes to pass a given point. The balloon at Banks' Ford is continually up. Long trains of wagons are still moving to the right.

T. S. C. LOWE.

11 A.M.

I can see no earth-works on the Bowling Green road. I should judge that the guns had been taken from the earth-works to the right of Fredericksburg. Another train of wagons is moving to the right on a road about one mile from beyond the heights opposite Franklin's Crossing. The enemy's barracks opposite Banks' Ford are entirely deserted. The largest column of the enemy is moving on the road toward Chancellorsville. The enemy on the opposite heights I judge considerably diminished. Can see no change under the heights and in the rifle-pits. I can see no diminution in the enemy's tents.

T. S. C. LOWE.

Professor T.S.C. Lowe's Official Report

12.30 P.M.

In a west-northwest direction, about twelve miles, an engagement is going on. Can see heavy smoke and hear artillery. In a west-southwest direction, about four miles, artillery is moving toward the engagement. A large force of the enemy are now digging rifle-pits-extending from Deep Run to down beyond the lower crossing just by the edge of woods at the foot of the opposite heights. There are but few troops in sight now except those manning batteries and in the rifle-pits. There appears to be a strong force in the rifle-pits.

T. S. C. LOWE.

2.15 P. M.

The enemy opposite here remain the same as last reported. Immense volumes of smoke are arising where the battle is going on opposite United States Ford. A large force must be engaged on both sides. This would be a good time for some staff officer to ascend, if it is desirable to you.

T. S. C. LOWE.

2.45 P. M.

The enemy are throwing up earth-works for artillery on a little rise of ground at the foot of the height about 300 yards from Deep Run.

T. S. C. LOWE.

Professor T.S.C. Lowe's Official Report

3.45 P. M.

The smoke from the battle appears to be in the same position, but in much lighter volumes. Everything opposite here remains the same.

T. S. C. LOWE,
Chief of Aeronautics, Army of the Potomac.

HEADQUARTERS ARMY OF THE POTOMAC,
May 2, 1863--5.15 *a.m.*

Professor LOWE:

Please get up your balloon at once and let me know the position of the enemy's troops.

DANL. BUTTERFIELD,
Major-General and Chief of Staff.

HEADQUARTERS ARMY OF THE POTOMAC,
May 2, 1863.

Professor LOWE:

Add to former dispatch and notice any movements toward Sedgwick's.

D. BUTTERFIELD,
Major-general.

Professor T.S.C. Lowe's Official Report

GENERAL SEDGWICK'S COMMAND,
May 2, 1863--6.15 *a.m.*

Major-General BUTTERFIELD, &C.:

The troops opposite this place remain in the same position as last evening. Owing to the high wind now prevailing I am unable to use a glass sufficiently to see whether there is any movement on the roads between here and the battleground of yesterday or not. I will examine them the first opportunity and report.

Respectfully, &c.,
T. S. C. LOWE.

MAY 2, 1863--7.80 *A. M.*

I have just obtained a tolerably good view of all the main roads beyond the heights and toward Chancellorsville, but could see no troops or wagon trains on them. The enemy opposite remain in the same positions, apparently without any increase.

Respectfully, &c.,
T. S. C. LOWE.

MAY 2, 1863--7.45 *A. M.*

General BUTTERFIELD, &C.:

Heavy cannonading has just commenced in a westerly direction about twelve miles. The enemy are shelling our troops opposite here.

Professor T.S.C. Lowe's Official Report

T. S. C. LOWE.

MAY 2, 1863--8.15 *A. M.*

Professor LOWE:

Has the enemy's force decreased any?

DANL. BUTTERFIELD,
Major-General and Chief of Staff.

8.30 A. M.

I cannot say that the enemy have decreased, but they do not show themselves quite as much this morning, and I can see no reserves on the opposite heights.

T. S. C. LOWE.

MAY 2, 1863--12 *M.*

Professor LOWE:

Why is not the balloon up, and why do we not hear from it?

DANL. BUTTERFIELD,
Major-General.

Professor T.S.C. Lowe's Official Report

12.30 P. M.

Major-General BUTTERFIELD, &C.:

GENERAL: I have made several efforts to ascend, but found the wind too high and could not use the glass. It is getting calmer now, and I will try again.

T. S. C. LOWE.

MAY 2, 1863--1.05 *P. M.*

The enemy remain the same opposite this point, and no movement is visible on any of the roads seen from the balloon. The wind continues so flawy that the balloon was blown from a thousand feet elevation to near the earth.

T. S. C. LOWE.

3.15 P.M.

A brigade of the enemy left from opposite the upper crossing fifteen minutes since, and crossed Deep Run, and is now moving to the right toward Banks' Ford. They have also disappeared from opposite our extreme left, below the lower crossing.

T. S. C. LOWE.

3.45 P.M.

Professor T.S.C. Lowe's Official Report

The enemy's troops that I saw moving to the right took the Plank road in the rear of Fredericksburg.

T. S. C. LOWE.

MAY 2, 1863--4.15 *P.M.*

The enemy have entirely withdrawn their advanced line, with the exception of a small picket force.

T. S. C. LOWE.

5.30 P.M.

Nearly all of the enemy's force have been withdrawn from the opposite side. I can only see a small force in the neighborhood of their earth-works. I cannot at this time get a sufficient elevation to tell what roads they take, but should judge by the appearance of army wagons moving to the right that the troops are moving that way also.

T. S. C. LOWE.

SIGNAL STATION, *May* 3, 1863.

Professor LOWE:

I am directed to inform you that your reports can be forwarded to headquarters Army of Potomac by telegraph. The station is where it was yesterday. Your reports to General

Professor T.S.C. Lowe's Official Report

Sedgwick can be forwarded by flag signals from station on bluff, immediately in front.

With great respect, your obedient servant,
F. WILSON,
First Lieutenant, in Charge Telegraph Station.

At 6 a.m. I was called upon by an aide, who said the general desired me to make a close examination of the enemy's position, and to point out his strongest and weakest points along the line of earthworks about Fredericksburg. The following was my report:

MAY 3, 1863--5.15 *A.M.*

Major-General SEDGWICK, *and*
General BUTTERFIELD,
Chief of Staff:

The enemy have apparently increased their force during the night, and appear again at the foot of the opposite heights. There does not appear to be as many, however, as yesterday morning.

T. S. C. LOWE.

7.15 A.M.

Major-General SEDGWICK, *and*
General BUTTERFIELD,
Chief of Staff:

The enemy's infantry is very light along the whole line opposite here, and especially immediately in the rear of Fredericksburg. I can see no troops moving this way on any of the

Professor T.S.C. Lowe's Official Report

roads. Heavy cannonading has just commenced on the right toward Chancellorsville.

T. S. C. LOWE.

Our troops were immediately concentrated in front, and at 11 o'clock the point reported by me to be the weakest was charged and very handsomely taken. I do not believe that any other point could have been taken by the same number of men.

HEADQUARTERS ARMY OF THE POTOMAC,
May 6, 1863--12.20 *p.m.*

Professor LOWE, *&c.:*

The commanding general wishes to have one balloon sent to United States Ford, inflated if possible. What answer shall I make to the general?

Very respectfully, your obedient servant,
S. WILLIAMS,
Assistant Adjutant-General.

In answer to the above dispatch I informed the general that I had but two balloons fit for use, one at Banks' Ford and the other at Fredericksburg, and that I would move whichever one to United States Ford he should direct. As it was necessary to know what movements the enemy were making in their rear, and the two places mentioned being the best for observations for that purpose, the general returned the following order:

UNITED STATES FORD, *May* 6, 1863.

General WILLIAMS:

Professor T.S.C. Lowe's Official Report

Leave the balloons for the present where they are--Fredericksburg and Banks' Ford.

J. HOOKER,
Major-General.

MAY 4, 1863 -12 *M.*

Generals SEDGWICK and BUTTERFIELD:

The enemy that entered the earth-works in the rear of Fredericksburg still remains. They also have considerable infantry and some wagons with their artillery on the heights to the left of Hazel Run. A portion of General Sedgwick's command occupies a position to the right commanding the enemy. I should estimate the enemy to be now in sight at least 15,000 strong.

T. S. C. LOWE.

6.15 P.M.

Generals HOOKER and SEDGWICK:

The enemy are advancing in large force to attack our forces on the right of Fredericksburg.

6.50 P.M.

The enemy are engaged in full force and driving our forces badly.

Professor T.S.C. Lowe's Official Report

MAY 4, 1863--7.30 P.M.

The enemy have driven our left with a large force and have possession of the river opposite Falmouth.

T. S. C. LOWE.

MAY 5, 1863--10.45 A.M.

Major-General BUTTERFIELD,
Chief of Staff:

I am unable at this time to see any movements of the enemy except some wagons moving up and some down the river. The enemy in force appear to hold all the ground they gained yesterday.

T. S. C. LOWE.

CAMP NEAR FALMOUTH, VA., May 7, 1863.

Capt. C. B. COMSTOCK,
Chief Engineer, Army of the Potomac:

CAPTAIN: The heavy storm of the 5th and 6th instant caused the loss of the entire gas from one balloon, partially from the other; also destroyed ten carboys of acid and four barrels of iron trimmings.

I would therefore respectfully recommend that 100 carboys of acid and twenty barrels of iron be at once ordered by telegraph.

I remain, very respectfully, &c.
T. S. C. LOWE,
Chief of Aeronautics, Army of the Potomac.

Professor T.S.C. Lowe's Official Report

Shortly after sending the above to Captain Comstock I called on him personally, relative to putting in order several balloons which needed repairs, and also to learn what decision had been made relative to my communication of April 12, 1863. Captain Comstock informed me that he would select the person to superintend that business---(the delicate one of putting balloons in order.) He also informed me that the terms were indicated in his indorsement on my communication. I informed him that was not satisfactory, and inasmuch as I had given notice on the 12th of April that I could not serve on the terms he named, and as the battle was now over, I wished to be relieved, provided it was a suitable time; to which Captain Comstock replied that if I was going I could probably be spared better then than any other time. I received pay up to April 7 inclusive, and came to Washington.

On the 8th I received the following dispatch, which is an indication that General Hooker was not informed of the change that had been made in the aeronautic department.

HEADQUARTERS ARMY OF THE POTOMAC,
May 8, 1863.

General Hooker sent one of his aides over at 10 a.m. to tell you to have two balloons up, and to keep them up all the time. I informed the aide that you had left the Army of the Potomac. Will you not write Hooker?

J. F. GIBSON.

Professor T.S.C. Lowe's Official Report

REPORT - CONCLUSION.

I have endeavored in this report not only to furnish a complete account of my own operations in connection with the military service, but to present all the essential facts for the use of the historian of this war relative to the introduction, use, and results of aeronautic observations. I feel assured that whatever may be the estimate of my own services, it will redound to the honor and credit of President Lincoln and his Administration that they have availed themselves of every means to crush this rebellion which loyal minds could devise or loyal men be willing to execute.

The details I have presented all have their significance when taken in connection with other facts relative to the conduct of the war known to the military authorities; and I have on this account, as well as from the entire novelty of the history, not thought it advisable to condense or abridge this report to a greater extent.

In conclusion, I would briefly state a few of the most important matters which deserve consideration.

First. The Government decided to introduce my system of aeronautics into the service---only after satisfactory experiments and practical tests had proved its importance--and it has been continued in constant use for two years under various generals, which would not have been the case had not experience demonstrated its utility, and the truth of all I originally claimed for it.

Second. Without wishing to disparage others, I may safely claim that my improved balloons and apparatus, including the portable gas generator (which are entirely my invention), are the

Professor T.S.C. Lowe's Official Report

only ones which are found to be adapted to the wants of the army service, and that I have done more to perfect the system, and to render it efficient and reliable than all who have been engaged in the art since the experiments of Guy Lussac in 1784.

To gain this knowledge has cost me many years hard labor and nearly $30,000 in money, and for which the United States Government alone is daily reaping the benefits.

Third. During the whole period of my employment I have devoted all my mental and physical energies to secure the success of the enterprise. I have never shrunk from the discharge of my duty, however hazardous, and holding no commission, I have often been perplexed and put to inconvenience in doing the business of the aeronautic department, which properly belonged to a commissioned officer, but for want of one acquainted with the business was compelled to do it myself. I have also been at all times exposed to the danger of being treated as a spy had I fallen into the hands of the enemy.

Fourth. For the first large balloon and apparatus brought to Washington and used in the preliminary experiments for the Secretary of War, and afterward at Falls Church, Fort Corcoran, and other places, I have never received compensation, nor for my labor and time, and expense of keeping a party of men employed for making the aforesaid experiments.

Fifth. It must be evident, without attempt at demonstration, that so novel and peculiar an apparatus as a balloon requires the most careful and trustworthy management and the most skilled and experienced observers. Having had more practical knowledge and greater experience in this business than any one else, I respectfully submit that the compensation I have asked and received has been small in comparison with the services performed.

Professor T.S.C. Lowe's Official Report

Sixth. The plans I have proposed are calculated to be of great value to the Army, and if proper facilities were afforded most important results could be obtained. Instead therefore of any curtailment of the aeronautic department I would most respectfully recommend its being permanently adopted as an arm of the military service, with established positions and regulations for those connected with its management. The persons to be selected for this service should be those tested in the field, and found to be the most reliable and experienced, who can instruct others when required. For want of proper facilities and persons capable for the service I have been unable to propose an extension of the balloon service to all parts of this army.

I have the honor to be, very respectfully, your most obedient servant,
T. S. C. LOWE,
Aeronaut.

P. S.--Since completing this report I have obtained a copy of the Prince de Joinville's narrative of the Peninsular Campaign, from which I extract the following:

Page 47:
The shells from the rifled guns flew in all directions with a length of range which had not before been suspected. The accuracy of their fire forced us to abandon all the signal posts we had established in the tops of the tallest trees. The balloon itself, whenever it rose in the air, was saluted with an iron hail of missiles which were, however, perfectly harmless.

Page 67:
Could the Federals meet, with a powerful concentration of troops, that concentration which the enemy had effected, and to the reality of which the observations of our aeronauts, as well as the statements of deserters, daily bore witness?

Professor T.S.C. Lowe's Official Report

Page 72:
It had been built by General Sumner, about half-way between Bottom's Bridge and the most advanced point of the Federal lines. It saved that day the whole Federal army from destruction.

NOTE.--I have the best of reasons to believe that Sumner's Bridge was completed a day sooner than it otherwise would have been by my frequent reports that the enemy were moving to the left. (See my dispatches to General McClellan of June 29, 1862, and following with comments.)

Page 75:
Some time had been lost under the impression that the attack on the right bank might be a feint. An end was soon put to all doubts on the subject by the vehemence of the attack, and by the aeronauts who reported the whole Confederate army moving to the scene of action. It was then that Sumner received the order to pass the river with his divisions.

NOTE.---See my dispatches of May 31 and June 1 with comments.

Page 86:
The presence of Jackson at Hanover Court-House proved that he intended to attack our communications, and cut them off by seizing the York River Railway. The maneuver was soon put beyond a doubt. A considerable body of troops were seen to leave Richmond, move in the direction of Jackson, and execute that movement to turn us, the danger of which we have already pointed out.

NOTE.--The above information was given in my reports of the 26th and 27th of June, 1862.

T. S. C. LOWE.

Professor T.S.C. Lowe's Official Report

Professor T.S.C. Lowe's Personal account of Operations During the Peninsula Campaign, 1862

It was through the midnight observations with one of my war-balloons that I was enabled to discover that the fortifications at Yorktown were being evacuated, and at my request General Heintzelman made a trip with me that he might confirm the truth of my discovery. The entire great fortress was ablaze with bonfires, and the greatest activity prevailed, which was not visible except from the balloon. At first the general was puzzled on seeing more wagons entering the forts than were going out, but when I called his attention to the fact that the ingoing wagons were light and moved rapidly (the wheels being visible as they passed each campfire), while the outgoing wagons were heavily loaded and moved slowly, there was no longer any doubt as to the object of the Confederates. General Heintzelman then accompanied me to General McClellan's headquarters for a consultation, while I, with orderlies, aroused other quietly sleeping corps commanders in time to put our whole army in motion in the very early hours of the morning, so that we were enabled to overtake the Confederate army at Williamsburg, an easy day's march beyond Yorktown on the road to Richmond.

Firing the day before had started early in the morning and continued until dark, every gun in the fortification being turned on the balloon, and then the next morning they were still pointing upward in the hope of preventing us in some way from further annoying the Confederates by watching their movements. The last shot, fired after dark, came into General Heintzelman's camp and

Professor T.S.C. Lowe's Official Report

completely destroyed his telegraph tent and instruments, the operator having just gone out to deliver a dispatch. The general and I were sitting together, discussing the probable reasons for the unusual effort to destroy the balloon, when we were both covered with what appeared to be tons of earth, which a great 12-inch shell had thrown up. Fortunately, it did not explode. I suggested that the next morning we should move the balloon so as to draw the foe's fire in another direction, but the general said that he could stand it if I could. Besides, he would like to have me near by, as be enjoyed going up occasionally himself. He told me that, while I saw a grand spectacle by watching the discharge of all those great guns that were paying their entire compliments to a single man, it was nothing as compared with the sight I would look down upon the next day when our great mortar batteries would open their siege-guns on the fortifications, which General McClellan expected to do.

I could see readily that I could be of no service at Williamsburg, both armies being hidden in a great forest. Therefore, General McClellan at the close of the battle sent orders to me to proceed with my outfit, including all the balloons, gas-generators, the balloon-inflating boat, gunboat, and tug up the Pamunkey River, until I reached White House and the bridge crossing the historic river, and join the army which would be there as soon as myself.

This I did, starting early the next morning, passing by the great cotton-bale fortifications on the York River, and soon into the little winding but easily navigated stream of the Pamunkey. Every now and then I would let the balloon go up to view the surrounding country, and over the bridge beyond the Pamunkey River valley, I saw the rear of the retreating Confederates, which showed me that our army had not gotten along as fast as it was expected, and I could occasionally see a few scouts on horseback on the hills beyond. I saw my helpless condition without my gunboat, the *Coeur de Lion,* which bad served me for the past year so well on the Potomac, Chesapeake, and York, and which I had sent to Commodore Wilkes to aid him in the bombardment of Fort Darling, on the James River, thinking I would have no further use

Professor T.S.C. Lowe's Official Report

for it. Therefore, all I had was the balloon-boat and the steam-tug and one hundred and fifty men with muskets, a large number of wagons and gas-generators for three independent balloon outfits. My balloon-boat was almost a facsimile of our first little Monitor and about its size, and with the flag which I kept at the stern it had the appearance of an armed craft, which I think is all that saved me and my command, for the *Monitor* was what the Confederates dreaded at that time more than anything else.

After General Stoneman had left me at White House. I soon had a gas-generating apparatus beside a little pool of water, and from it extracted hydrogen enough in an hour to take both the general and myself to an altitude that enabled us to look into the windows of the city of Richmond and view its surroundings, and we saw what was left of the troops that bad left Yorktown encamped about the city.

While my illness at Malvern Hill prevented me from reporting to headquarters until the army reached Antietam. those in charge of transportation in Washington took all my wagons and horses and left my command without transportation. Consequently I could render no service there, but the moment General McClellan saw me he expressed his regret that I had been so ill, and that he did not have the benefit of my services; for if he had he could have gotten the proper information, he could have prevented a great amount of stores and artillery from recrossing the Potomac and thus depleted the Confederate army that much more. I explained to him why he had been deprived of my services, which did not surprise him, because be stated that everything bad been done to annoy him, but that be must still perform his duty regardless of annoyances. When I asked him if I should accompany him across the river in pursuit of Lee, he replied that he would see that I had my supply trains immediately, but that the troops after so long a march were nearly all barefoot, and in no condition to proceed until they bad been properly shod and clothed.

Without the time and knowledge gained by the midnight observations referred to at the beginning of this chapter, there would have been no battle of Williamsburg, and McClellan would have lost the opportunity of gaining a victory, the importance of

Professor T.S.C. Lowe's Official Report

which has never been properly appreciated. The Confederates would have gotten away with all their stores and ammunition without injury. It was also my night observations that gave the primary knowledge which saved the Federal army at the battle of Fair Oaks.

On arriving in sight of Richmond, I took observations to ascertain the best location for crossing the Chickahominy River. The one selected was where the Grapevine, or Sumner, Bridge was afterward built across that stream. Mechanicsville was the point nearest to Richmond, being only about four miles from the capital, but there we would have bad to face the gathering army of the Confederacy, at the only point properly provided with trenches and earthworks. Here I established one of my aeronautic stations, where I could better estimate the increase of the Confederate army and observe their various movements. My main station and personal camp was on Gaines' Hill, overlooking the bridge where our army was to cross.

When this bridge was completed, about half of our army crossed over on the Richmond side of the river, the remainder delaying for a while to protect our transportation supplies and railway facilities. In the mean time, the Confederate camp in and about Richmond grew larger every day.

My night-and-day observations convinced me that with the great army then assembled in and about Richmond we were too late to gain a victory, which a short time before was within our grasp. In the mean time, desperate efforts were made by the Confederates to destroy my balloon at Mechanicsville, in order to prevent my observing their movements.

At one point they masked twelve of their best rifle-cannon, and while taking an early morning observation, all the twelve guns were simultaneously discharged at short range, some of the shells passing through the rigging of the balloon and nearly all bursting not more than two hundred feet beyond me, showing that through spies they had gotten my base of operations and range perfectly. I changed my base, and they never came so near destroying the balloon or capturing me after that.

I felt that it was important to take thorough observations that

Professor T.S.C. Lowe's Official Report

very night at that point, which I did. The great camps about Richmond were ablaze with fires. I had then experience enough to know what this meant, that they were cooking rations preparatory to moving. I knew that this movement must be against that portion of the army then across the river. At daylight the next morning, May 31st, I took another observation, continuing the same until the sun lighted up the roads. The atmosphere was perfectly clear. I knew exactly where to look for their line of march, and soon discovered one, then two. and then three columns of troops with artillery and ammunition wagons moving toward the position occupied by General Heintzelman's command.

All this information was conveyed to the commanding general, who, on hearing my report that the force at both ends of the bridge was too slim to finish it that morning, immediately sent more men to work on it.

I used the balloon *Washington* at Mechanicsville for observations, until the Confederate army was within four or five miles of our lines. I then telegraphed my assistants to inflate the large balloon, *Intrepid,* in case anything should happen to either of the other two. This order was quickly carried out, and I then took a six-mile ride on horseback to my camp on Gaines Hill, and made another observation from the balloon *Constitution.* I found it necessary to double the altitude usually sufficient for observations in order to overlook forests and hills, and thus better to observe the movements of both our army and that of the Confederates.

To carry my telegraph apparatus, wires, and cables to this higher elevation, the lifting force of the *Constitution* proved to be too weak. It was then that I was put to my wits' end as to how I could best save an hour's time, which was the most important and precious hour of all my experience in the army. As I saw the two armies coming nearer and nearer together, there was no time to be lost. It flashed through my mind that if I could only get the gas that was in the smaller balloon, *Constitution,* into the *Intrepid,* which was then half filled, I would save an hour's time, and to us that hour's time would be worth a million dollars a minute. But how was I to rig up the proper connection between the balloons? To do this within the space of time necessary puzzled me until I

Professor T.S.C. Lowe's Official Report

glanced down and saw a 10-inch camp kettle, which instantly gave me the key to the situation. I ordered the bottom cut out of the kettle, the *Intrepid* disconnected with the gas-generating apparatus, and the *Constitution* brought down the hill. In the course of five or six minutes connection was made between both balloons and the gas in the *Constitution* was transferred into *the Intrepid*.

I immediately took a high-altitude observation as rapidly as possible, wrote my most important dispatch to the commanding general on my way down, and I dictated it to my expert telegraph operator. Then with the telegraph cable and instruments, I ascended to the height desired and remained there almost constantly during the battle, keeping the wires hot with information.

The Confederate skirmish line soon came in contact with our outposts, and I saw their whole well-laid plan. They had massed the bulk of their artillery and troops, not only with the intention of cutting off our ammunition supplies, but of preventing the main portion of the army from crossing the bridge to join Heintzelman.

As I reported the movements and maneuvers of the Confederates, I could see, in a very few moments, that our army was maneuvering to offset their plans.

At about twelve o'clock, the whole lines of both armies were in deadly conflict. Ours not only held its line firmly, but repulsed the foe at all his weaker points.

It was one of the greatest strains upon my nerves that I ever have experienced, to observe for many hours a fierce battle, while waiting for the bridge connecting the two armies to be completed. This fortunately was accomplished and our first reenforcements, under Sumner, were able to cross at four o'clock in the afternoon, followed by ammunition wagons.

It was at that time that the first and only Confederate balloon was used during the war. This balloon, which I afterward captured, was described by General Longstreet as follows:

" It may be of interest at the outset to relate an incident which illustrates the pinched condition of the Confederacy even

Professor T.S.C. Lowe's Official Report

as early as 1862.

 The Federals had been using balloons in examining our positions, and we watched with envious eyes their beautiful observations as they floated high up in the air, well out of range of our guns. While we were longing for the balloons that poverty denied us, a genius arose for the occasion and suggested that we send out and gather silk dresses in the Confederacy and make a balloon. It was done, and we soon had a great patchwork ship of many varied lines which was ready for use in the Seven Days campaign.

 We had no gas except in Richmond, and it was the custom to inflate the balloon there, tie it securely to an engine, and run it down the York River Railroad to any point at which we desired to send it up. One day it was on a steamer down on the James River, when the tide went out and left the vessel and balloon high and dry on a bar. The Federals gathered it in, and with it the last silk dress in the Confederacy. This capture was the meanest trick of the war and one that I have never yet forgiven.
"

UNITED STATES.

Professor T.S.C. Lowe's Official Report

Fitz John Porter Views The Confederates From A Balloon

On the 11th of April [1862] at five o'clock, an event at once amusing and thrilling occurred at our quarters. The commander-in-chief had appointed his personal and confidential friend, General Fitz John Porter, to conduct the siege of Yorktown. Porter was a polite, soldierly gentleman, and a native of New Hampshire, who had been in the regular army since early manhood. He fought gallantly in the Mexican war, being thrice promoted and once seriously wounded, and he was now forty years of age, handsome, enthusiastic, ambitious, and popular. He made frequent ascension with Lowe, and learned to go aloft alone. One day he ascended thrice, and finally seemed as cosily at home in the firmament as upon the solid earth. It is needless to say that he grew careless, and on this particular morning leaped into the car. and demanded the cables to be let out with all speed. I saw with some surprise that the flurried assistants were sending up the great straining canvas with a single rope attached. The enormous bag was only partially inflated, and the loose folds opened and shut with a crack like that of a musket. Noisily, fitfully, the yellow mass rose into the sky, the basket rocking like a feather in the zephyr; and just as I turned aside to speak to a comrade, a sound came from overhead, like the explosion of a shell, and something striking me across the face laid me flat upon the ground.

Half blind and stunned, I staggered to my feet, but the air seemed full of cries and curses. Opening my eyes ruefully, I saw all faces turned upwards, and when I looked above, the balloon was adrift.

The treacherous cable, rotted with vitriol, had snapped in twain; one fragment had been the cause of my downfall, and the other trailed, like a great entrails from the receding car, where Fitz John Porter was bounding upward upon a Pegasus that he could neither check nor direct.

Professor T.S.C. Lowe's Official Report

The whole army was agitated by the unwonted occurrence. From battery No. 1, on the brink of the York, to the mouth of Warwick river, every soldier and officer was absorbed. Far within the Confederate lines the confusion extended. We heard the enemy's alarm-guns, and directly the signal flags were waving up and down our front.

The General appeared directly over the edge of the car. He was tossing his hands frightenedly, and shouting something that we could not comprehend.

"O-pen-the-valve! " called Lowe, in his shrill tones; "climb-to-the-netting-and-reach-the-valve-rope."

"The valve!-the valve!" repeated a multitude of tongues, and all gazed with thrilling interest at the retreating hulk that still kept straight upward, swerving neither to the east nor the west.

It was a weird spectacle,-that frail, fading oval, gliding against the sky, floating in the serene azure, the little vessel swinging silently beneath, and a hundred thousand martial men watching the loss of their brother in arms, but powerless to relieve or recover him. Had Fitz John Porter been drifting down the rapids of Niagara, he could not have been so far from human assistance. But we saw him directly, no bigger than a child's toy, clambering up the netting and reaching for the cord.

"He can't do it," muttered a man beside me; "the wind blows the valverope to and fro, and only a spry, cool-headed fellow can catch it."

We saw the General descend, and appearing again over the edge of the basket, he seemed to be motioning to the breathless hordes below, the story of his failure. Then he dropped out of sight, and when we next saw him, he ,as reconnoitering the Confederate works through a long black spy-glass. A gloat laugh went up and down the lines as this cool procedure was observed, aid then a cheer of applause ran from group to group. For a moment it was doubtful that the balloon would float in either direction; it seemed to falter, like an irresolute being, and moved reluctantly southeastward, towards Fortress Monroe. A huzza, half uttered, quivered on every lip. All eyes glistened, and some were dim with tears of joy. But the wayward canvas now turned due westward, and was blown rapidly toward the Confederate works. Its course was fitfully direct, and the wind seemed to veer often, as if contrary currents, conscious of the opportunity, were struggling for the possession of the daring navigator. The south wind held mastery for awhile, and the balloon passed the Federal front amid a howl of despair from the soldiery. It kept right on, over sharpshooters, rifle-pits, and outworks, and finally passed, as if to deliver up its freight, directly over the heights

Professor T.S.C. Lowe's Official Report

of Yorktown.

The cool courage, either of heroism or despair, had seized upon Fitz John Porter. He turned his black glass upon the ramparts and masked cannon below, upon the remote camps, upon the beleaguered town, upon the guns of Gloucester Point, and upon distant Norfolk. Had he been reconnoitering from a secure perch at the tip of the moon, he could not have been more vigilant, and the Confederates probably thought this some Yankee device to peer, into their sanctuary in despite of ball or shell. None of their great guns could be brought to bear upon the balloon; but there were some discharges of musketry that appeared to have no effect, and finally even these demonstrations ceased. Both armies in solemn silence were gazing aloft, while the imperturbable mariner continued to spy out the land.

The sun was now rising behind us, and roseate rays struggled up to the zenith, like the arcs made by showery bombs. They threw a hazy atmosphere upon the balloon, and the light shone through the network like the sun through the ribs of the skeleton ship in the *Ancient Mariner.* Then, as afl looked agape, the air-craft "plunged, and tacked, and veered," and drifted rapidly toward the Federal lines again.

The allelujah that now went up shook the spheres, and when he had regained our camp limits, the General was seen clambering up again to clutch the valve-rope. This time he was successful, and the balloon fell like a stone, so that all hearts once more leaped up, and the cheers were hushed. Cavalry rode pell-mell from several directions, to reach the place of descent, and the General's personal staff galloped past me like the wind, to be the first at his debarkation. I followed the throng of soldiery with due haste, and came up to the horsemen in a few minutes. The balloon had struck a canvas tent with great violence, felling it as by a bolt, and the General, unharmed, had disentangled himself from innumerable folds of oiled canvas, and was now the cynosure of an immense group of people. While the officers shook his hands, the rabble bawled their satisfaction in hurrahs, and a band of music marching up directly, the throng on foot and horse gave him a vociferous escort to his quarters.

Townsend - "Campaigns of a Non-Combatant"

www.ingramcontent.com/pod-product-compliance
Lightning Source LLC
Chambersburg PA
CBHW071418160426
43195CB00013B/1730